RAND NATIONAL DEFENSE RESEARCH INSTITUTE

Research-Portfolio Performance Metrics

Rapid Review

Marjory S. Blumenthal, Jirka Taylor, Erin N. Leidy, Brent Anderson, Diana Gehlhaus Carew, John Bordeaux, Michael G. Shanley

Prepared for the Office of the Secretary of Defense

For more information on this publication, visit **www.rand.org/t/RR2370**

Library of Congress Cataloging-in-Publication Data is available for this publication.
ISBN: 978-1-9774-0162-5

Published by the RAND Corporation, Santa Monica, Calif.

Cover: Based on image from Adobe Stock / vectorfusionart.

Preface

This report is the final deliverable for a project involving rapid review of research-portfolio performance metrics. The overarching objective of the study was to undertake a review of methods and metrics to assess research progress at the level of research portfolios, with a view to informing performance-based decisions at various levels of leadership and management. The request for this study came from what was then known as the Defense Centers of Excellence for Psychological Health and Traumatic Brain Injury (DCoE) (now part of the Defense Health Agency). It took, as a point of departure, federal agencies' need to comply with the reporting requirements of the Government Performance and Results Act (Pub. L. 103-62, 1993) and the Government Performance and Results Act Modernization Act (Pub. L. 111-352, 2011), which proved to offer broader context than specific guidance for the activities documented in the report. The project, although it took note of DCoE's areas of focus and context as part of the Military Health System, included broad consultations across other research organizations (prioritizing agencies that were involved at least partially with medical research), as well as wider literature on evaluating research portfolios. The resulting findings are relevant to the Defense Health Agency (especially its Research and Development Directorate) and others involved in research sponsored by the U.S. Department of Defense and supporting the readiness, health, and well-being of the warfighter, and they are likely to be of interest to other supporters of research, given the comparative compilations presented.

This research was sponsored by what was then known as DCoE and conducted within the Forces and Resources Policy Center of the RAND National Defense Research Institute, a federally funded research and development center sponsored by the Office of the Secretary of Defense, the Joint Staff, the Unified Combatant Commands, the Navy, the Marine Corps, the defense agencies, and the defense Intelligence Community.

For more information on the RAND Forces and Resources Policy Center, see www.rand.org/nsrd/ndri/centers/frp or contact the director (contact information is provided on the webpage).

Contents

APPENDIXES

Figures

Tables

Summary

Federal agencies (and others) that support research are expected to evaluate the effectiveness of their research-related programs, just as they would evaluate performance of other kinds of programs. Legislative frameworks, such as the Government Performance and Results Act of 1993 (Pub. L. 103-62), as updated in the Government Performance and Results Act Modernization Act of 2010 (Pub. L. 111-352, 2011), make sure that attention to such evaluation is paid at the highest levels of an agency, although responsibility for gathering data lies at lower levels, closer to where the work is done. Compared with some kinds of activities, such as procuring materiel or delivering services, research can be hard to evaluate; accordingly, research evaluation differs from program evaluation. Challenges include lags between the time research is completed and when its outcomes and impacts can be observed, as well as the attribution of any observed results to the underpinning research. Through both formal study and practical experience, people responsible for research administration and research evaluation have developed an array of metrics that can help in evaluating research performance at different levels, from the individual project through different levels of aggregation.

In requesting RAND Corporation assistance in understanding how others evaluate research performance, what was then the Defense Centers of Excellence in Psychological Health and Traumatic Brain Injury (DCoE) (now part of the Defense Health Agency) sought insight into measurement of the performance of research portfolios—aggregates of research projects and sometimes research programs that are themselves composed of projects. The portfolio level is holistic, internalizing the heterogeneity and complementarity among projects that are typical of the volume of activity found at an agency level, in particular. To facilitate learning about research-portfolio evaluation, the primary research question addressed in this report is, what methods and metrics do organizations use to assess research performance at the level of the research portfolio? The analysts for this study developed a database of research-portfolio metrics through a rapid review—a time-constrained scan of relevant published materials and series of consultations with people responsible for research evaluation in a set of federal and some private agencies and organizations. The 34 entities from which the rapid review drew include agencies and organizations most prominent and renowned in terms of the development, execution, and evaluation of portfolios of a variety of types of research—basic, applied, and translational.

The RAND team developed a taxonomy of research-portfolio metrics that was guided by a logic model–based approach. A logic model is a conceptualization of individual program components (inputs, processes, outputs, outcomes, and impacts) and their relationships. It expresses a theory of how research efforts are expected to lead to desired results. Selecting a logic-model approach as the conceptual basis for this study provided the research team with clear principles for organizing collected data in a way that is broadly comparable across various contexts. The team organized a database of identified metrics in line with the main components of the logic model, with individual metrics categorized into broader metric types and aggregate metric categories.

The team found that, broadly speaking, research organizations use three types of portfolio-level metrics: (1) aggregations of project-level data, derived by adding up data from individual projects; (2) narrative assessments; and (3) general (e.g., population-level) metrics, which might or might not depend on project-level data. Each of these three types of portfolio-level metrics has its advantages and disadvantages. For instance, although aggregations of project metrics tend to be relatively easy to compile and communicate, they can lack important nuance. By contrast, narrative assessments are well suited to describing the results of the underlying research, as well as its unique contribution, but they can be costly and not amenable to comparisons across contexts. General (e.g., population-level) metrics tend to build on available data and are easily understandable, but they can give rise to substantial issues attributing causality for any measured results. Consequently, the interests of research-portfolio assessment are best served by a mixed-methods approach balancing multiple types of metrics.

Further, informed by the data collected through the document and literature review, as well as stakeholder interviews, we observed a greater effort outside of the components we examined from the U.S. Department of Defense (DoD) than inside it expended on measuring outputs and the ultimate consequences of research-portfolio investments—outcomes and impacts—than on the perhaps more–easily measured inputs and processes (see Figure S.1). We observed a similar pattern in data compiled in the database of metrics and variables used in the tracking system developed and piloted by DCoE in 2017.

The review found that there is no one-size-fits-all approach to research-portfolio assessment, a finding in line with existing literature. Practices observed across research organizations, as well as organizations' conceptualizations of portfolios, vary accordingly. During the

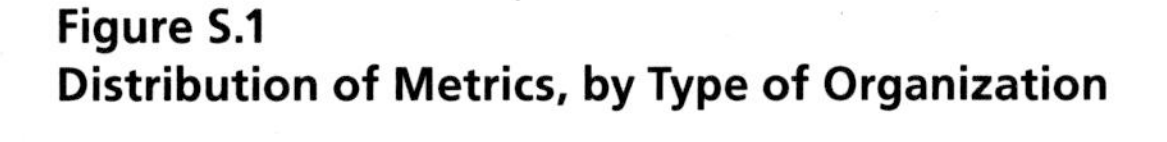

Figure S.1
Distribution of Metrics, by Type of Organization

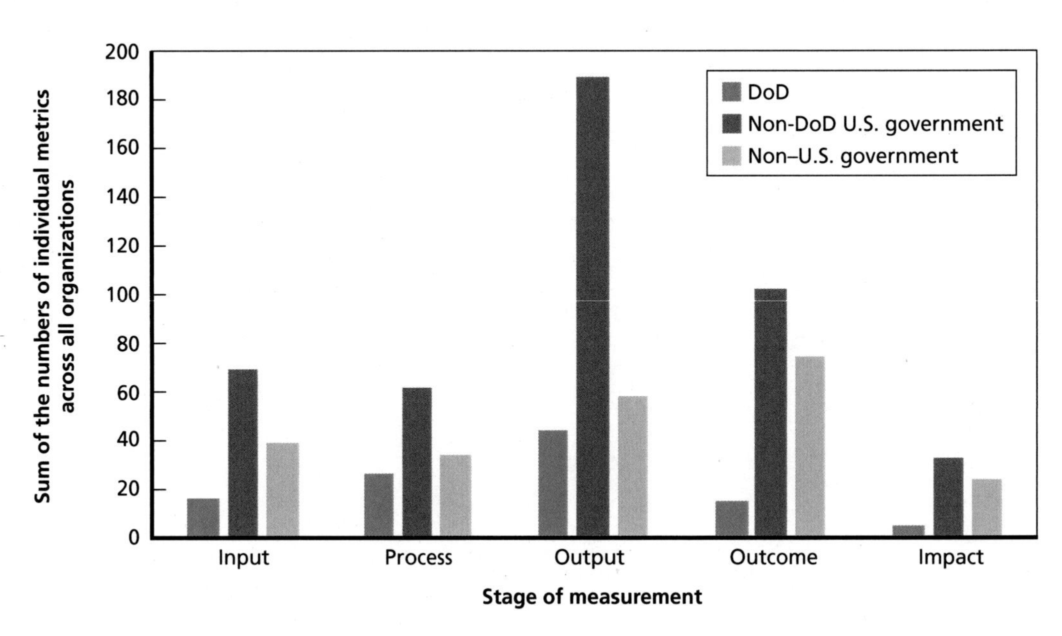

review, we also observed that noteworthy innovative work in this area is taking place across research organizations. These efforts take multiple forms, ranging from setting up dedicated infrastructure to assess research portfolios to building evaluation capacity and systems, as well as developing new tools. These developments suggest strong potential for cross-organizational learning (which was part of the motivation for this project).

The collected set of research-portfolio metrics, which illuminates both common practice and occasional innovation by sponsors of research portfolios, and the commentary provided by both literature and interviews were the basis for the research team's deliberations and our judgments in formulating a series of recommendations. These are primarily intended for the former DCoE and its successors, now subsumed by the Defense Health Agency. They might, however, be applicable more broadly, particularly to other DoD entities that share a common set of goals—support for the health, well-being, and readiness of warfighters—and a comparatively applied and translational focus for their research. The recommendations are as follows:

- **Review the value of currently collected data on upstream metrics (inputs and processes).** The objective of such a review would be to ascertain the continued utility of the current scope of data collection and whether the benefits of all collected data outweigh the costs associated with their collection. This recommendation arises from the observation that the current scope of upstream data collection is extensive, whereas less is done with downstream metrics (especially outcomes and impacts), for which resources could be used more effectively.
- **Identify opportunities for streamlining reporting requirements and activities.** The objective of this activity would be to review whether there could be greater efficiencies in existing reporting arrangements and data management. This recommendation arises from concerns expressed by interviewees in DoD entities about reporting burdens and the positive examples gleaned from non-DoD entities of the use of central information systems.
- **Incorporate outcome and impact measurements in tracking and assessment processes.** The development of an impact measurement framework can be an important first step in making an assessment of outcomes and impacts more robust. This recommendation arises from two observations: First, non-DoD entities have been doing more than the DoD entities that we examined to measure research-portfolio outcomes and impacts; and second, among the DoD entities examined, there is an opportunity for measurement of research-portfolio outcomes and impacts that is more systematic, which appears, from communication with the sponsor, to be a goal within DoD.
- **Consider developing outcome and impact tracking and measurement in an incremental fashion.** It might not be feasible to simultaneously introduce a broad suite of outcome and impact metrics. Instead, their gradual implementation, focusing initially on a small number of selected metrics, might be more realistic. This recommendation reflects our judgment that wholesale change might not be practical and that experimentation with alternatives (informed by the examples collected and discussed in this report) could be beneficial.
- **Construct a balanced mix of metrics and determine how underlying data will be collected.** The selection of metrics for outcome and impact measurement needs to consider the trade-offs associated with their use and balance the efforts needed to collect data to inform these metrics with their utility to key stakeholders, as well as their intended

use. This recommendation reflects lessons learned from the literature and our judgment that, in a world of research constraints and performance-measurement demands, there is an opportunity to make explicit choices about the metrics used at each stage represented by the logic model. These choices might achieve a better balance than what was observed and reported here without adding to reporting burdens.

Acknowledgments

The research team is grateful for the interest in and support for this project from its sponsors, beginning with Dennis Goodes, who had the vision to request the study, and his current and former associates Bryan Kahner, Catherine Haight, Jennifer Weil, William Klenke, Ada Determan, and Andrew Smith. Coming late to the oversight of the project, CAPT Joseph Cohn brought well timed enthusiasm for the value and utility of the project's results.

In Appendix E, we list the many people involved with research portfolios at federal and some private agencies and organizations who contributed input about their practices, experiences, and insights. We could not have done this research without their generosity and collegiality.

The team appreciated the support and process guidance from RAND colleagues Kristie L. Gore, John D. Winkler, Craig A. Bond, Barbara Hennessey, Sarah O. Meadows, and Neil DeWeese. We benefited from the expertise of Charles C. Engel, Sarah Parks, and Daniel Basco.

Finally, we are grateful for the perspectives provided by the reviewers engaged through the RAND Corporation quality assurance process. Their thoughtful feedback helped immeasurably in strengthening this report.

Abbreviations

AHRQ	Agency for Healthcare Research and Quality
CAHS	Canadian Academy of Health Sciences
CDC	Centers for Disease Control and Prevention
CDMRP	Congressionally Directed Medical Research Programs
CEP	Commission on Evidence-Based Policymaking
CoE	center of excellence
DCOE	Defense Centers of Excellence for Psychological Health and Traumatic Brain Injury
DHA	Defense Health Agency
DoD	U.S. Department of Defense
DVBIC	Defense and Veterans Brain Injury Center
ERA	Excellence in Research for Australia
FDA	U.S. Food and Drug Administration
FY	fiscal year
GAO	U.S. Government Accountability Office
GPRA	Government Performance and Results Act of 1993
GPRAMA	Government Performance and Results Act Modernization Act of 2010
HHS	U.S. Department of Health and Human Services
IRB	institutional review board
IRIS	Institute for Research on Innovation and Science
J-9	Research and Development Directorate
JPC-1	Joint Program Committee 1 (Medical Simulation and Information Sciences)
MHS	Military Health System

NAS	National Academy of Sciences
NASA	National Aeronautics and Space Administration
NIDILRR	National Institute on Disability, Independent Living, and Rehabilitation Research
NIH	National Institutes of Health
NIOSH	National Institute for Occupational Safety and Health
NIST	National Institute of Standards and Technology
NRAP	National Research Action Plan
NSF	National Science Foundation
OECD	Organisation for Economic Co-operation and Development
OMB	Office of Management and Budget
OPA	Office of Portfolio Analysis
PART	Program Assessment Rating Tool
PH	psychological health
PHCoE	Psychological Health Center of Excellence
PI	principal investigator
PSI	Population Services International
PTSD	post-traumatic stress disorder
RCR	Relative Citation Ratio
REF	Research Excellence Framework
RePORTER	Research Portfolio Online Reporting Tools
ROI	return on investment
RPM	Research Portfolio Management
SUD	substance-use disorder
T2	National Center for Telehealth and Technology
TBI	traumatic brain injury
UK	United Kingdom
VA	U.S. Department of Veterans Affairs

CHAPTER ONE

Introduction

The important economic and societal gains from research, including achievements made through federally funded research programs, have been well documented (Murphy and Topel, 2003; Hall, Mairesse, and Mohnen, 2010; National Research Council, 2014). Existing research-evaluation literature documents numerous potential approaches to research assessment (Brutscher, Wooding, and Grant, 2008; Grant, Brutscher, et al., 2010; Grant and Wooding, 2010; Wilsdon et al., 2015; Milat, Bauman, and Redman, 2015; Greenhalgh et al., 2016). This introductory chapter sets the context by briefly discussing objectives of efforts to evaluate research generally, describing the evolution of initiatives to evaluate the performance of federal agency activities that provide additional motivation for evaluating federally supported research, and outlining considerations surrounding research evaluation at the level of the research portfolio, an evaluation level encouraged by those federal initiatives. The chapter closes by presenting the objectives of this study, which aimed to collect and understand metrics used to evaluate research portfolios, along with our methodology.

As noted in existing literature (see, e.g., Morgan Jones and Grant, 2013; Hinrichs-Krapels and Grant, 2016; Ovseiko et al., 2016), research evaluation and assessments are typically motivated by at least one of four drivers, conceptually organized as the four As: accountability, analysis, allocation, and advocacy. Examining the performance of research organizations can illuminate each of these four As. With accountability, the objectives are to demonstrate how effectively and efficiently investments in research have been used and to hold researchers accountable to research managers and funders. For analysis, the objective is to collect information on and improve understanding of why and how research efforts are effective and how further research support can be provided. An allocation objective aims to help inform future allocation of resources. And last, an advocacy objective could take several forms, including (1) demonstrating the merits of investments in research, (2) promoting the understanding of research among policymakers and the general public, and (3) supporting the case for changes in policy and practice (Guthrie, Wamae, et al., 2013).

For this project, the authors of this report aimed to enhance evaluation of research that is comparatively applied and focused on health and health care, although the work benefited from considering evaluation of a much wider range of research. The project team focused on research evaluation at the level of the portfolio, as opposed to individual projects or even groups of projects pursued as programs when they themselves are components of larger portfolios that are more complex. In general, the research portfolios that were the basis of inputs to the project were found, as opposed to constructed—the research supported by an agency composed its research portfolio. As discussed in the next two sections and throughout the report, a portfolio perspective presents both measurement challenges and opportunities to address the four As in new and specific ways.

Federal Government Performance Measurement Initiatives

Measuring how well agencies have been performing their missions has been encouraged by a series of federal government initiatives that provide context and motivation for this study. Although they do not focus explicitly on research activities, their provisions have, inter alia, applied to agencies with substantial research portfolios. They inform agency leadership interests about ways to measure the performance of research portfolios and in understanding how different agencies use such measurements or metrics (terms that are used interchangeably in this report). In particular, they foster a focus on the portfolio level of evaluation (discussed in the next section) because, as described below, they drive agency-level measurement and reporting.

In an effort to improve performance oversight across the federal government, Congress passed the Government Performance and Results Act of 1993 (GPRA) (Pub. L. 103-62) and the GPRA Modernization Act of 2010 (GPRAMA) (Pub. L. 111-352, 2011). These laws mandate that each agency articulate performance goals and objectives for itself each year, against which its performance is then measured.[1] Accordingly, agencies and departments that conduct research or award federal research grants must annually evaluate their federally funded research programs and portfolios. These evaluations are typically done at a high level (i.e., organization- or agency-wide) within the federal agency or department. They help to motivate federal agencies that fund research to collect data and track each funded program and project throughout its life cycle, including after the research is completed. Moreover, agencies are required to compile and estimate metrics on these research portfolios and report them in an informative, systematic way.

In fulfilling their obligations under GPRAMA, individual agencies are invited to conduct their own evaluations and reporting. As the National Academy of Sciences (NAS) observed in a 2001 study on the implementation of GPRA, "given the diversity in mission, complexity, culture, and structure of federal agencies that support research, it is not surprising that their approaches to GPRA have varied. One size definitely does not fit all" (Institute of Medicine, NAS, and National Academy of Engineering, 2001). The U.S. Government Accountability Office (GAO), which has produced a series of monitoring reports on the act's implementation (see, e.g., GAO, 2015a; GAO, 2015b; and GAO, 2016), has encouraged information-sharing and exchange of best practices among agencies (see the box). Because GPRAMA operates at a very high level, supervised by staff associated with agency heads and aggregating a lot of information, it provides, at best, a backdrop for research-evaluation efforts at an agency.

Other initiatives at the federal level designed to help measure and improve performance include the short-lived Program Assessment Rating Tool (PART) and the Commission on Evidence-Based Policymaking (CEP). PART was introduced by the Office of Management

[1] Examples of strategic documents that present performance objectives and goals in line with GPRA and GPRAMA requirements include the National Science Foundation's (NSF's) strategic plan for fiscal years (FYs) 2018 through 2022 (NSF, 2018) or the U.S. Department of Health and Human Services' (HHS's) strategic plan for FYs 2014 through 2018 (HHS, 2018).

Implementation of GPRAMA

The implementation of GPRAMA has been subject to GAO monitoring and reporting. In 2015, GAO published a review of the act's implementation, having found that agencies continued to face issues using performance information, although the report noted gradual implementation of processes, such as reviews of agency priority goals, which could lead to improvements in this area (GAO, 2015b). The 2015 review also highlighted that agencies continued to encounter difficulties with respect to connecting agency performance and results, as well as collecting and reporting financial and performance data in selected areas.

Also in 2015, GAO published an assessment of how selected agencies reported on the quality of performance information they use to track progress toward their organizational performance goals, as stipulated by GPRAMA (GAO, 2015a). The review found that the agencies under review generally did not make publicly available the information on how they ascertained the accuracy and reliability of their performance data.

and Budget (OMB) in 2003 and mandated that each major federal government program assess its performance in the following areas: (1) program purpose and design, (2) strategic planning, (3) program management, and (4) program results (Gilmour, 2007). PART was discontinued in 2009 and followed by GPRAMA, with the objective of further promoting the use of performance information.[2] Research suggests that neither engagement with GPRA performance reporting processes nor engagement with PART assessments had much direct effect on the use of performance information as part of agency management routines (Moynihan and Lavertu, 2012). One exception to this observation was the use of performance information for the refinement of agency measures and goals, which was found to have been encouraged by GPRA and PART.

In 2016, the congressionally mandated bipartisan CEP was established to examine possibilities to enhance the availability and use of data with the objective of building evidence to inform the design of policy programs. This includes an exploration of ways to integrate and make available existing administrative and survey data to support research and evaluation activities and possible steps to enhance existing data infrastructure (OMB, 2016). In September 2017, the commission published its findings, to which we return in pertinent parts of this report (CEP, 2017).

More recently, the latest relevant GAO report as of this writing is aptly titled *Government-Wide Action Needed to Improve Use of Performance Information in Decision Making* (GAO, 2018). It ties critique of limited reporting of performance information to the March 2018 President's Management Agenda (President's Management Council and Executive Office of the President, 2018), which highlights cross-agency priority goal (required by GPRAMA), such as "leveraging data as a strategic asset" (President's Management Council and Executive Office of the President, 2018, p. 17).

Another noteworthy example of federal efforts in this area has been the STAR METRICS initiative led by the National Institutes of Health (NIH) and NSF, intended to develop tools and collect data that could help evaluate the impact of federal research spending. As of 2016, the initiative's resources had been directed toward supporting the development of Federal Research

[2] Even after the discontinuation of PART, OMB remains relevant for the purposes of federal agencies' performance measurement, not least via its circulars. For instance, OMB Circular A-11 stipulates how agencies' performance goals should be incorporated into their budget requests and makes an explicit reference to GPRA and GPRAMA (OMB, 2017). The same circular contains instructions on federal laboratories' annual reporting in the area of technology transfer, which feeds into National Institute of Standards and Technology (NIST)–prepared annual federal laboratory technology transfer reports (see, e.g., NIST, 2018a).

Portfolio Online Reporting Tools (RePORTER) (STAR METRICS, 2015). In Chapter Two, we discuss this data repository.[3]

Why Evaluate Portfolios

The focus for research evaluation can range from individual researchers or research projects (sometimes referred to as *studies*) to programs or institutions, to the entirety of a research field or national research system. Against that backdrop, there is no universally accepted definition of *research portfolio*, and, as evidenced by stakeholders' testimony collected for this study, the term can be meaningfully applied to any level of analysis above individual projects and researchers. To illustrate the multiple nomenclature possibilities within the Military Health System (MHS), a frequent conceptualization of a portfolio is that of a body of research addressing a given medical issue, such as traumatic brain injury (TBI). TBI could correspond to a research field, although the overall TBI-field research portfolio could be disaggregated in numerous ways—for instance, by a particular aspect (e.g., by various stages on the continuum of care) or by particular organizations sponsoring or conducting research relevant for this field. Moreover, as Guthrie, Wamae, et al., 2013, highlights, the unit of analysis can differ with respect to data collection and data reporting. Data are typically collected at a granular level and subsequently reported either at the same level or aggregated to apply to a higher level of analysis.[4]

As noted above, portfolio evaluation is driven by legal reporting requirements, as well as by recommendations from a variety of experts and bodies concerned about good governance (see, e.g., CEP, 2017). Even without such external motivation, federal agencies awarding research funds, such as those associated with the Defense Health Agency (DHA), want to identify overlaps in projects and program areas, address gaps in topic funding relative to organizational goals, and understand the impact of funded research. The agencies want these to inform guidance for framing future budget requests from Congress and internal resource allocation (see, e.g., Wallace and Rafols, 2015).

Growing Interest Among Funders and Performers in Portfolio Evaluation

As numerous observers have pointed out, interest in the evaluation of research activities and portfolios has been on the rise (e.g., Srivastava, Towery, and Zuckerman, 2007; Organisation for Economic Co-operation and Development [OECD], 2009; Wallace and Rafols, 2015; Linton and Vonortas, 2015). In a 2015 report, analysts from the United Kingdom (UK) Independent Review of the Role of Metrics in Research Assessment and Management observed, "there are powerful currents whipping up the metric tide" (Wilsdon et al., 2015, p. viii). As that quotation illustrates, indicators used to measure the performance of research activities and portfolios are often called *metrics*, a term used in commissioning this project.

---

[3] Data collected as part of the STAR METRICS initiative also inform the work by the Institute for Research on Innovation and Science (IRIS), established by a consortium of universities and hosted by the University of Michigan. IRIS manages an information and analytical platform with longitudinal data on federally funded research in an effort to foster better understanding and analysis of research activities and their impacts (IRIS, undated).

[4] See Appendix A for an overview of data collection and data reporting levels of analysis used by selected international research-evaluation frameworks.

Several trends, consistent with the four As listed above, have been identified as contributing to the growing interest in research-portfolio evaluation. First, in the current era of tight budgets, Congress and other research funders are looking for ways to help them prioritize the allocation of their resources (OECD, 2009). Competition for scarce resources, such as funding, students, and staff, creates pressure for more assessments via rivalry between (as well as within) research institutions (Wilsdon et al., 2015). Second, research evaluation and assessments reflect growing demands from policymakers for data on research progress, impact, and quality, motivated by accountability considerations and by a desire to highlight the economic and societal gains from resources invested in research (OECD, 2009; Wilsdon et al., 2015). Finally, researchers, as well as decision makers, have increasing amounts of data available to inform their decisions (Hicks et al., 2015), and new data analytic tools enable evaluation to be more rigorous (Wallace and Rafols, 2015).

Portfolio Evaluations Can Take Place at Multiple Stages of the Research Process

Assessments and evaluations of research portfolios can be undertaken at various stages of the research process as embodied by the portfolio's components. Three broad types of assessments can be distinguished.

First, prospective evaluations (sometimes referred to as *impact assessments*) are conducted before the commencement of research activities. They typically help inform decisions about what research activities should be undertaken and how resources should be allocated. As part of this process, the likelihood of achieving desired impact, as well the magnitude of this impact, might be assessed.

Second, portfolio evaluations can be conducted while research activities are ongoing. Such assessments aim to assess the state of the research portfolio and progress toward its stated aims.

Third, retrospective evaluations are conducted after the conclusion of research activities within a portfolio. The objective of such evaluations is to reflect on the results of the research and what has been achieved with the research investments and efforts made.

The focus of this report is on the latter two types of evaluations. The first type, prospective evaluations, has been covered in existing literature primarily from the perspective of the prominent use of peer review as a means of assessing proposals for research funding (see, e.g., Guthrie, Guerin, et al., 2013; Chalmers et al., 2014; Li and Agha, 2015; Guthrie, Ghiga, and Wooding, 2017). Prospective impact assessments are limited and typically take the form of statements of envisaged impacts; those statements appear as part of funding applications (see, e.g., NSF, 2017). The more intentionally a portfolio is composed, the more easily it can be evaluated against the initial objectives (Linton and Vonortas, 2015). Discussions in this report focus on evaluating existing portfolios.

Objectives and Methodology of This Study

The research question for the project culminating in this report was, What methods and metrics do notable sponsors of research in the United States use to evaluate their portfolios of research? The result is a review of methods and metrics to assess portfolios during and after the component research is conducted. *Metric* is the term commonly used to refer to a way of measuring aspects of either research activity or research consequences (stages for doing so are discussed below). The working conceptualization of a portfolio adopted by the research team was that of a body of research bringing together multiple constituent parts (e.g., individual projects or studies [terms used interchangeably], programs into which individual projects fit and that themselves might fit into larger portfolios or those that are more complex) working toward a common goal. Through consultations undertaken as part of this project, the research team examined how various organizations and agencies conceived of their research portfolios and addressed challenges associated with evaluations and assessments conducted at that level of analysis. Although the study was requested by a part of the MHS that emphasized applied research relating to health and health care and the translation to clinical practice (MHS, undated), it drew on wider research-evaluation evidence where appropriate and instructive.

The project was designed to support a set of broad aims: (1) to facilitate compliance with various obligations to report on research performance, (2) to foster an understanding of how various performance metrics can be and are applied by various research organizations as an evidence base to inform their decision making, and (3) to help senior leaders in their decision making about how research portfolios contribute to the achievement of high-level objectives. High-level objectives for the project sponsor are clear: improvement and maintenance of military personnel health, well-being, and readiness.

Conceptual Approach

Given the scope of the project, which includes a review of portfolio metrics applicable to both research that is ongoing and that has been completed, we used an approach based on logic models as the study's organizing principle (for a discussion of logic models, see, e.g., Greenfield, Williams, and Eiseman, 2006; Taylor-Powell and Henert, 2008; Landree, Miyake, and Greenfield, 2015; and Savitz, Matthews, and Weilant, 2017). A logic model (sometimes referred to as a *logical framework*) represents graphically the key components or elements of a program and its consequences. These elements are typically categorized into inputs, activities, outputs, outcomes, and impacts. Logic models illuminate individual activity components and how these various components relate to each other. The process of developing a logic model requires formulating a theory of how a program's activities can be plausibly expected to lead to its envisaged or desired results. As such, a logic model facilitates a systematic examination of both the individual elements of a program and any underlying assumptions and external factors that might play a role in determining whether the ultimate outcomes and impacts are achieved.[5]

For the purposes of this project, the use of a logic model–based analytical approach brings two principal advantages. First, it offers a conceptually clear way of organizing collected data that can be applied across various contexts and institutions. Second, examining existing practices through the prism of a logic model enables a better understanding and comparison of what evidence research organizations consider important to collect at individual stages of their result chains.

Figure 1.1 depicts a simple example of a logic model as applied to research. Although it shows a linear flow, the relationship between individual components is considerably more complex than that. The purpose of the logic-model format is to provide a simplified way of

---

[5] For additional literature on the topic, see, e.g., Coryn et al., 2011; Guthrie, Wamae, et al., 2013; Milat, Bauman, and Redman, 2015; and Greenhalgh et al., 2016.

Figure 1.1
Illustrative Graphical Representation of a Logic Model

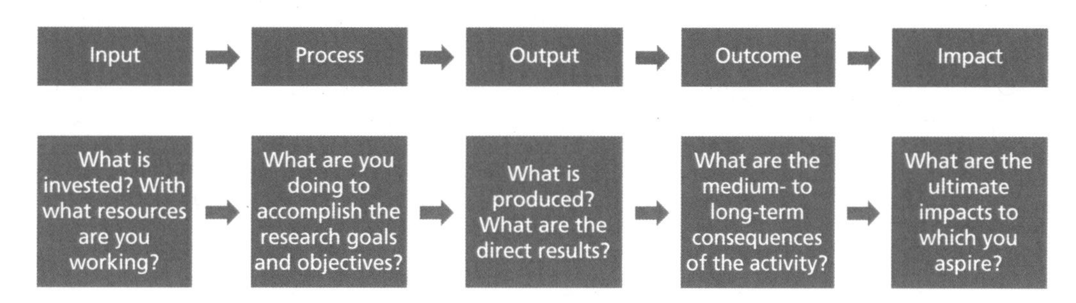

thinking about the various steps that are expected to ultimately feed into expected impacts. The individual components of a logic model as used in this study are

- **inputs:** *Inputs* generally refers to resources (human, material, financial), as well as mechanisms (e.g., existing relationships, networks, panels), that are invested in the research.
- **processes:** *Processes* refers to steps undertaken to produce research findings (e.g., research projects, methods), individuals and entities involved in the research, and their mode of contribution (e.g., collaborations, study populations)
- **outputs:** *Outputs* refers to direct, tangible products of research activities. These can come in various types and are typically quantifiable. Examples include published papers and reports, new methods, and knowledge generated.
- **outcomes and impacts:** This category encompasses the intended (as well as unintended) results of the research activities, typically linked to research objectives. In some contexts, making a firm distinction between outcomes and impacts might not be meaningful or possible, although, as we attempted here, one possibility is to differentiate outcomes from impacts by assigning different time frames to these two categories. Under this approach, outcomes would be short- and medium-term results from research, typically observable before long-term results (i.e., impacts) manifest themselves. According to this conceptualization, outcomes represent intermediate results (e.g., changes in clinical practice, guidelines, or policy informed by research), which are expected to lead to desired ultimate results, or impacts (e.g., improvements in health achieved via the changes in practice or policy).

Research evaluation can occur at every stage of the research cycle (see, e.g., Wilsdon et al., 2015). Reflecting the individual steps in a logic model, research-evaluation metrics can be applied to inputs, processes, outputs, outcomes, and impacts. Previous literature has commented on the suitability of metrics at individual stages of measurement (see, e.g., Guthrie, Wamae, et al., 2013). Metrics in the earlier stages of the logic model (upstream measures) tend to be used for accountability and advocacy purposes. In contrast, downstream measures, such as outcomes and impacts, tend to be associated with learning and steering objectives. One reason for this is the existence of time lags between the conduct of research and the materialization of its impacts (Brutscher, Wooding, and Grant, 2008; Canadian Academy of Health Sciences [CAHS], 2009).

Methods

To draw on rigorous research approaches while adhering to a comparatively brief time frame, we adopted a rapid review approach to develop an understanding of performance tracking and associated metrics across research organizations. Although there is no universally valid definition of *rapid review methodology* (see, e.g., Khangura, Polisena, et al., 2014), it is generally understood to refer to a method of evidence synthesis based on an adaptation of elements of a systematic review and designed to produce results in a constrained period of time (see, e.g., Khangura, Danko, et al., 2012; Tricco et al., 2015).[6]

We operationalized the rapid review via a two-pronged approach to data collection (document review and key-informant interviews), followed by internal synthesis and assessment. In the rest of this section, we describe each of these three stages in greater detail.

Entities We Covered in Our Review

Data collection drew primarily on the organizations and agencies listed in Table 1.1, from each of which at least one metric was identified. As a starting point in identifying agencies and organizations for review, the research team followed the mapping of stakeholders utilized in a financial analysis for the former Defense Centers of Excellence for Psychological Health and Traumatic Brain Injury (DCoE) conducted in parallel to this study (unpublished) to identify suitable organizations for review. Using a snowball approach, we identified additional agencies and organizations, largely following up on recommendations from stakeholders consulted for the study.

Table 1.1 clusters entities by whether they are part of the U.S. Department of Defense (DoD), are other federal agencies, or are nongovernmental organizations. The review team prioritized agencies and other organizations that are involved either wholly or at least partially with medical research. Examples of the latter are agencies that conduct or fund research in a variety of areas, of which health and health care are one. This group of agencies includes the National Aeronautics and Space Administration (NASA), NIST, and NSF. Involvement in applied health or health care research was welcome but was not a prerequisite for inclusion in the study sample. In consultation with the sponsor, we regularly updated the selection of agencies for review. Although the number consulted is limited, the group includes leaders and exemplars in research support and evaluation. Please note that the group of agencies listed in the table combines multiple levels of analysis. For instance, U.S. Army Medical Research and Materiel Command subsumes other entities listed among DoD stakeholders. Similar overlaps can be observed among non-DoD stakeholders—several organizations are part of HHS. The inclusion of multiple levels of analysis in this review reflects the fact that portfolio assessments are meaningful to and undertaken by all these entities.

Literature Review

The objective of this activity was to identify metrics used by various organizations in assessing the progress and results of their research. We organized the data collection into three strands. First, we conducted a general and rapid search of existing literature on research evaluation and performance assessments, with special focus on publications authored by federal government agencies, given the original motivation provided by GPRA and GPRAMA compliance.

---

[6] For an overview of various types of methods employed in rapid reviews, as well as their possible limitations and biases, see a review by Ganann, Ciliska, and Thomas, 2010.

Table 1.1
Overview of Agencies Included in the Review

Category	Agency
DoD	• CDMRP • Extremity Trauma and Amputation CoE • Hearing CoE • Medical Simulation and Information Sciences Research Program • Military Operational Medicine Research Program • Naval Health Research Center • Uniformed Services University of the Health Sciences • U.S. Air Force 59th Medical Wing • U.S. Army Medical Research and Materiel Command • Vision CoE • Walter Reed Army Institute of Research
Non-DoD U.S. government	• AHRQ • CDC • FDA • HHS • NASA • National Institute of Environmental Health Sciences • National Occupational Research Agenda • NIDILRR • NIH • NIOSH • NIST • NSF • VA
Non–U.S. government	• Association of American Medical Colleges • Bill and Melinda Gates Foundation • Howard Hughes Medical Institute • Kaiser Permanente Washington Health Research Institute • Mathematica Center on Health Care Effectiveness • Patient-Centered Outcomes Research Institute • Pennington Biomedical Research Center • PSI[a] • Researchfish[b] • Robert Wood Johnson Foundation

NOTE: CDMRP = Congressionally Directed Medical Research Programs. CoE = center of excellence. AHRQ = Agency for Healthcare Research and Quality. CDC = Centers for Disease Control and Prevention. FDA = U.S. Food and Drug Administration. NIDILRR = National Institute on Disability, Independent Living, and Rehabilitation Research. NIOSH = National Institute for Occupational Safety and Health. VA = U.S. Department of Veterans Affairs. PSI = Population Services International.

[a] PSI is a nongovernmental organization that aims to improve the availability of health produces in services in low- and middle-income countries (PSI, undated).

[b] Researchfish is a research-impact assessment used by a large number of research funders in Europe and North America, including all of the UK's Research Councils. Its member organizations include the Bill and Melinda Gates Foundation through its Grand Challenges Explorations and the Wellcome Trust (Researchfish, undated). For a discussion of the platform, see, e.g., Hinrichs, Montague, and Grant, 2015.

Second, we reviewed the websites of agencies and organizations selected for review. And third, we conducted a series of targeted online searches (e.g., using Google Advanced Search, probing the Defense Technical Information Center) utilizing a series of key search terms.[7]

To ensure consistency in data collection across the research team, we used a standardized Excel data-extraction template that formed the basis for the eventual database of identified

[7] These were *research evaluation, research assessment, research portfolio, portfolio review, research performance, performance evaluation, performance assessment, research metric, performance metric, portfolio metric, research indicator, performance indicator, annual report, progress report, financial report, performance appendix, performance plan, strategic goal,* and *strategic plan.*

metrics. The template helped research team members to log any identified metrics and document any relevant information associated with these metrics, such as units of measurement, frequency of use, and limitations, where offered by the reviewed literature. Our data collection captured both metrics that the agency or organization claimed or documented as having used and metrics that have been proposed for use by a given agency.

Stakeholder Interviews

The second component of our methodology was a series of stakeholder interviews to learn about approaches to research-portfolio assessment in their respective organizations.[8] Information collected through the interviews also complemented the identification of research-portfolio metrics undertaken through document review. The selection of agencies and organizations to consult mirrored the selection approach for the document review. We used an opportunistic approach to the sampling of interviewees, starting with aspirational invitations to selected individuals and following up on any recommendations and suggestions received. As a general rule (albeit with exceptions), for DoD stakeholders, we drew on DCoE's recommendations for suitable interviewees and followed up on introductions made by DCoE representatives. For stakeholders from non-DoD agencies, we reached out either to individuals identified through the document review or to members of the research team's existing professional networks. We offered invited key informants to nominate other people from their respective organizations if they did not feel comfortable commenting on the subject of this study.

The interviews were semistructured and conducted either by phone or in person. They followed a standardized topic guide structured around the objectives and research questions for this study (provided in Appendix D). This approach ensured a degree of consistency across all interviews while allowing for a discussion of unique and unanticipated topics. We advised interviewees that their responses and statements would not be attributed to them. The research team conducted 30 interviews with a total of 42 interviewees.

Synthesis of Metric Data

Following the collection of individual metrics from reviewed agencies, the research team developed a categorization system to enable an aggregate analysis of the data, as well as cross-organizational comparisons, because the context and operationalization of even similar metrics can differ across agencies. This categorization system consisted of organizing identified metrics at four distinct analytical levels (see Figure 1.2).

As the first step, we categorized individual metrics by the stages of measurement corresponding to the five components of the logic model (analytical level 1). We categorized these at the collection stage as part of the data-extraction template and subsequently had two members of the research team review each categorization independently. Within each stage of measurement, we grouped individual metrics into broad metric categories (analytical level 2). For instance, we categorized as "staff" any metric pertaining to human resources invested in research. As the next step, within each broad metric category, we identified concrete metric types that would bring together similar or identical individual metrics across all organizations (analytical level 3). For instance, within the "staff" category of metrics, we identified metric types, such as "indicators of researcher quality." In some instances, the individual metrics

[8] The study protocol was submitted for review to the Human Subjects Protection Committee, RAND's Institutional Review Board (IRB). The study was categorized in August 2017 as not involving human subjects.

Figure 1.2
Approach to the Categorization of Identified Metrics

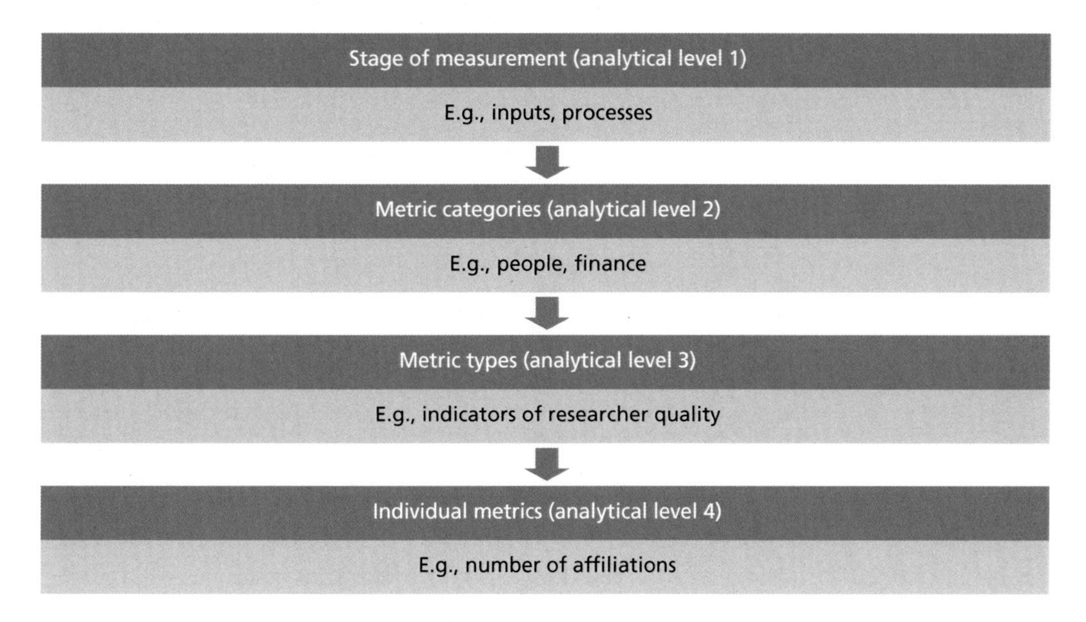

(which represent analytical level 4) grouped under each metric type were identical or nearly identical across all relevant agencies. To illustrate, there is no variation across agencies within the metric type of "number of researchers" (i.e., each individual metric of this metric type is identical). In other instances, e.g., the metric type "indicators of researcher quality," there is some variation in the individual metrics used by organizations. For instance, some organizations might look at the seniority of funded researchers, while others might assess their various professional affiliations.

The analysis presented in this report used primarily *metric types* (i.e., *analytical level 3*) as the unit of analysis to accommodate the large number of individual metrics identified. We are confident that this level of aggregation allows for a meaningful and approachable presentation of findings. Where there are notable instances of variation within individual metric types, we discuss these in the relevant sections.

Assessment of Metric Data

To process collected evidence, the project team developed a framework to assess the types of metrics identified in this review. This framework, captured in Table 1.2, was designed by building on existing literature in the fields of research evaluation (e.g., Guthrie, Wamae, et al., 2013) and performance management (e.g., Paul, Yeats, Clark, and Matthews, 2015; Paul, Yeats, Clark, Matthews, and Skrabala, 2015; Savitz, Willis, et al., 2015). We also drew on a toolkit on measures for policymakers, developed by Savitz, Matthews, and Weilant and presented at a series of workshops for representatives of the Office of the Under Secretary of Defense for Policy focusing on measurement challenges in policies and programs (Savitz, Matthews, and Weilant, 2017).

The framework incorporates three assessment criteria that we judged relevant for the objectives of this report: reliance on expert judgment, data burden, and attribution or contribution issues. These three criteria loosely correspond to the dimensions of (statistical) reliability,

Table 1.2
Metric Assessment Criteria

Value	Reliance on Expert Judgment	Data Burden	Attribution Issues
Low	The metric makes use of data that require little expert input.	The required data are available and take relatively little effort to collect.	The metric is well suited to capture research contribution.
Medium	The metric requires some expert judgment.	The required data are available but take substantial effort to collect (e.g., from external organizations, after the end of research activities).	The metric is an indirect proxy for a measure of interest.
High	The metric relies primarily on expert judgment, frequently in the form of qualitative assessments.	The required data would be challenging to collect.	The metric has substantial attribution or contribution challenges.

feasibility, and validity, as discussed in existing literature.[9] Within each criterion, we assessed each metric type on a three-point scale, ranging from low to high. Importantly, these assessments should not be understood as conveying desirability. Specifically, in our conceptualization, reliance on expert judgment broadly corresponds to the dimension of reliability in its statistical meaning (i.e., whether assessments yield consistent results under repeated measurements). It would be incorrect, for example, to conclude that a metric type with a low degree of reliance on expert judgment can be considered inferior (or superior) to one incorporating expert judgment to a greater extent. Similarly, although (all else being equal) types of metrics with low data burden might be preferable to those with high data burden, they might be less desirable from a variety of other perspectives. At the same time, benefits to having the information available could still outweigh any considerations related to data-collection costs.

Ultimately, the objective of this assessment was to help illustrate the trade-offs associated with the selection and utilization of research metrics. For instance, as one interviewee pointed out, "things that are easiest to measure aren't always the most useful." Hence, a portfolio evaluation will be served best by a combination of metrics, drawing on a mix of tools, methods, and metrics.

The criterion of reliance on expert judgment captures the extent to which a particular metric requires consideration and input from various stakeholders. Metrics can range from primarily quantitative measures that are meaningful on their own and require little interpretation to those that incorporate substantial qualitative input (e.g., success narratives). No metric can be considered completely devoid of expert judgment. Even metrics that require little interpretation once the data are collected can require judgment with respect to definitions, for example. For such metrics, it is likely that expert judgment will be drawn on in earlier, design-oriented stages of measurement. To illustrate, although the number of high-quality publications can

[9] This report uses specific criterion labels instead of general terms, such as *reliability* and *validity*, in an effort to maintain clarity and avoid giving an impression of a normative judgment. We used the three criteria primarily to describe the qualities of individual metric types, not to convey any sense of the metrics' relative superiority or inferiority. Furthermore, of course, other criteria could be applied in assessing metrics. An example of another criterion is relevance or utility (i.e., a metric's usefulness to its users and other stakeholders) (see, e.g., Paul, Yeats, Clark, and Matthews, 2015). For instance, even if a metric is reliable, feasible, and valid, if it is completely unrelated to the aims of the program, it could be useless. Because the use of this criterion is context dependent, we decided not to include it in the assessment framework, which we intended to apply across various contexts.

be a relatively straightforward metric to collect and interpret, it requires consideration of what constitutes the threshold for a product to be considered high quality.[10]

The criterion of data burden captures how difficult it can be to collect data necessary to inform the metric in question. Data burden can be high in situations in which the specific type of information is not commonly available, as well as those in which the collection of information requires consulting a multitude of stakeholders and other data holders, particularly if these stakeholders are external to the organization conducting the portfolio assessment (e.g., end users of the research) and the data need to be collected after the underlying research activities have concluded. Indeed, time lags in research (discussed in greater detail in Chapter Two) can be among the biggest drivers of data-collection burdens. We heard frequently from interviewees, 13 of whom explicitly stressed resource constraints as a limiting factor, that data burden can grow with the frequency, volume, and variety of data requests, as well as with volatility in data requests. Four interviewees also remarked on reporting burden that researchers bear and the need to keep it to a minimum.[11]

The criterion of attribution issues reflects the fact that metrics vary substantially in their ability to capture the unique contribution that the underlying research made to the observed or claimed results. It expresses whether a given metric type can express the research's true contribution and minimize the uncertainty surrounding the possibility that (or degree to which) the observed or claimed results are a product of other possible factors.

In the process of constructing the database of metrics, we developed the scoring of individual metric types against individual criteria. The scoring was validated in the course of an internal synthesis workshop, in which the research team discussed the results of the review of metrics and reflected on the resulting findings and conclusions. An important overarching qualification is that this assessment represents an effort to develop a heuristic indicator of the relative characteristics of individual metrics and associated trade-offs. How metrics score on these three criteria in practice will inevitably depend on the mode of their use.

Analysis of Interview Data

Through the lens of thematic analysis, broadly consistent with the methods outlined in Braun and Clarke, 2006, and Fereday and Muir-Cochrane, 2006, we examined information collected via stakeholder interviews. In this analysis, we gathered the notes from all conducted interviews, drew out overarching themes from the texts, and then further organized these by idea. We began by manually looking for patterns and themes from the notes, based on answering the questions posed to the interviewees, utilizing an Excel spreadsheet. These themes included defining *portfolio*, the audience for metrics, how research topics are identified, cooperation with other organizations, challenges, best practices, discontinuation of projects, and sources of data. Once we identified broad themes, we systematically reviewed each interview for the respondent's comments on each topic. In addition, we tracked any metrics the interviewees

[10] Because the role of expert judgment in designing and applying quantitative measures might not be immediately obvious, those measures' end users might consider them "objective." This can give rise to challenges, such as gender bias, in bibliometric analyses.

[11] Of course, the amount of burden associated with data collection depends on the way the metric is used in practice and how it fits with the organization's other activities. For instance, if an organization already conducts a regular survey of its researchers, expanding the survey to include questions pertaining to portfolio assessments might cost relatively little. In the absence of an existing data-collection mechanism, the burden on organizations and researchers associated with establishing one can be substantial.

mentioned as being used in their organizations to add to the list of metrics identified from literature.

After we reviewed the interviews, we proceeded to organize the data within individual themes. In this step, we manually searched for patterns within each theme to find suitable organizing principles (or ideas) at a more granular level. For example, within the theme of "defining *portfolio*," responses could be grouped into bins corresponding to various ways of organizing portfolios. These included, for instance, considerations revolving around research topics, funding streams, or methods. For a topic to qualify as a stand-alone idea within a theme, at least two interviewees had to make an observation relevant to that topic. We captured in an "other" category any unique observations that only one interviewee mentioned. Once we finished classifying themes and ideas, we created frequency counts at both levels of analysis (i.e., by theme and idea) across all interviews.

Limitations of the Project

This project report is subject to several limitations, which we acknowledge up front. First, definitions of *portfolio* vary across, as well as within, agencies and organizations, especially between various levels of decision making. In addition, the context in which agencies undertake their research varies, as do the objectives of the research activities. This heterogeneity necessarily places limitations on how much evidence from individual agencies and organizations is comparable. To mitigate this issue, we undertook some degree of generalization while extracting, processing, and categorizing individual research metrics. In addition, in aggregating data on metrics, we selected an analytical level that we believe lends itself best to cross-contextual comparisons.

Second, the selection of agencies in this review was not exhaustive and did not include all possibly relevant agencies and organizations. That said, this review achieved a meaningful coverage of federal health research organizations. We examined or consulted organizations representing the majority of U.S. federal health research funding, and we covered principal research organizations that are not focused solely on health research (NASA, NIST, and NSF). The degree to which this review covers private health research efforts is difficult to assess.

Third, in a few instances, we did not secure an interview with a representative of an organization included in the review. In these cases, we report on metrics that are listed in the organizations' publicly available documents. That listing might be incomplete because the organization might use other metrics not described in public-facing documents. This might have been a particular challenge for DoD-sponsored research and related activities because we found only limited information available online. Furthermore, it might be that, even where we were able to consult with stakeholders, our conversations might have not touched on all conceivable metrics in use by the office, agency, or organization. Consequently, our review can only evidence the use of a particular metric, not document that an organization does *not* use a given metric.

Finally, the analysis was based on what is already in use (although it also drew on literature featuring suggestions of metrics whose actual use is uncertain). That a metric is used does not necessarily mean that it is "good" or appropriate; the absence of use does not mean that a metric lacks value or at least potential. A focus on what is being used is intrinsically more conservative than the potential landscape of metric and indicators.

In the introduction to Chapter Three, we discuss the practical implications of these limitations on our database of metrics and on the interpretation of the data.

Structure of This Report

The remainder of this report is structured as follows. Chapter Two discusses a series of general findings pertaining to portfolio metrics in use by reviewed organizations and a set of common challenges associated with their use. Chapter Three presents a detailed overview of identified metrics and our assessments of them. Chapter Four concludes the report by offering a set of broad, big-picture observations, as well as observations and recommendations for successors of DCoE.

Several appendixes complement the report. Appendix A summarizes additional insights from academic and practitioner literature regarding existing research-evaluation frameworks and tools. Appendix B offers a suggested prioritization of performance metrics, reflecting DCoE's Research Portfolio Management (RPM) office's data dictionary. Appendix C provides additional data on and breakdowns of metrics identified in the course of this rapid review. Appendix D reproduces the content of the protocol used during stakeholder interviews. Appendix E provides a list of stakeholders who informed this study and agreed to be acknowledged. Appendix F reproduces DCoE's RPM data dictionary.

CHAPTER TWO

General Findings and Considerations

Summary of Main Points

Broadly speaking, three types of portfolio-level assessments can be distinguished:

- aggregations of project-level metrics
- narrative portfolio assessments
- general (e.g., population-level) metrics.

Each of these three types of assessments has its own advantages and disadvantages, suggesting that the purposes of portfolio measurement might be best served by a mixed approach.

Existing literature, as well as stakeholder testimony, highlighted several principal challenges with respect to portfolio evaluation:

- Some research portfolios might be too heterogeneous to allow some forms of meaningful analysis and assessment at an aggregate level.
- The success of portfolio assessment depends on the quality and availability of underpinning data, which can be burdensome to ensure.
- Time lags in research can mean that a substantial period of time might pass before any expected outcomes and impacts materialize.
- It might be difficult to establish a clear link between the underpinning research and any claimed results and thus attribute the achieved outcomes and impacts to the original research efforts.

This chapter provides an overview of high-level findings and observations on the general context of portfolio performance measurement and associated challenges we found in our work. This chapter is organized in two main parts. First, we discuss broad types of portfolio-level metrics that research organizations employ, along with the trade-offs stemming from their use. Second, we discuss a series of key challenges associated with portfolio performance measurement highlighted by existing research-evaluation literature, as well as by interviewed stakeholders.

Forms of Portfolio-Level Metrics

Portfolio-level analysis typically involves some degree of aggregation of data collected at lower levels of analysis, although this aggregation is not always necessary. Our rapid review of existing literature, as well as stakeholder interviews, indicated that research organizations employ, broadly, three types of portfolio-level metrics. First is an aggregation of project-level metrics, typically derived by summing up data collected from individual projects or studies. Second is a narrative portfolio assessment, which utilizes primarily qualitative approaches to take stock of a given portfolio and its results. And third is general, frequently population-level, metrics that are quantitative in nature but do not always require making use of data collected at lower levels of analysis.

Each metric form has its advantages and disadvantages, which we discuss in turn in this chapter. It is important to note that these three types of metrics are not mutually exclusive and that organizations can (and do) use multiple types in a complementary way. As observed in Guthrie, Wamae, et al., 2013, the use of mixed methods might represent the most appropriate response to existing research-evaluation needs.

Four interviewees highlighted challenges associated with focusing on portfolio metrics. One interviewee noted that aggregation at the portfolio level can give rise to interpretation challenges or hide important information. It might be possible either to meet all required milestones and still not produce valid research or to have great research that is not useful for the portfolio in question. Another interviewee added that identifying metrics that would be applicable and meaningful at high levels of aggregation, particularly at the level of entire agencies (e.g., in line with GPRAMA requirements) is very difficult. Yet another interviewee opined that one challenge for portfolio metrics is to present data in a way that is clear and helpful for decision makers who face multiple demands on their time and do not have the capacity to explore the data and their nuances, even though aggregation at the portfolio level is likely to require caveats and explanations.

Aggregations of Project Metrics

The aggregation of project-level metrics typically takes the form of summing up the values reported for individual studies forming the portfolio in question. This type of aggregation is particularly well suited to "upstream" measures of inputs, processes, and outputs (as well as some outcomes) because these are typically readily quantifiable. For instance, an organization might assess the overall number of publications that its research portfolio has produced.

The summing approach to aggregation has several strengths. First and foremost, it is relatively easy to compile, as well as to communicate. With respect to the former, assuming that individual projects and studies report relevant data, the aggregation itself is very straightforward. With respect to the latter, aggregations of project-level metrics tend to be transparent and easy to understand. Furthermore, because of their relatively low burden and simplicity, they lend themselves well to various types of comparisons, including across organizations and over time.

At the same time, this approach has several disadvantages. Most importantly, aggregations of this type can lack nuance. Particularly with respect to comparisons, a certain degree of judgment or categorization might be necessary to produce measures that can be applied to every context. In addition, some forms of quantification can obscure important pieces of information or differences in the source data. Using the example of the number of publications, an aggregate sum would treat each item as equally important when, in reality, they are not. There are ways to mitigate this challenge (e.g., use quality measures alongside quantity ones, focus on peer-reviewed publications or select journals only), but their effectiveness is limited. Furthermore, this approach to aggregation largely does not attempt to capture the added value of research portfolios (i.e., it does not examine whether the observed achievements are larger than they would have been if the contributing projects were not part of one coherent research effort or program). (Of course, the identification of any portfolio's added value might not be part of the assessment's objectives.) Aggregation can also introduce the risk of double-counting, if multiple projects contribute to the same output or outcome.

Narrative Portfolio Assessments

Another approach to aggregate assessments is the use of qualitative narratives. The main strength of this approach is that it enables a nuanced and in-depth assessment of portfolio performance. In particular, narrative approaches can address attribution issues (discussed in more depth later in this section) and help clearly demonstrate the link between the research and the claimed results. This is particularly important for "downstream" metrics (outcomes and impacts), which are most susceptible to attribution issues.

A notable technique in this regard is the development of a contribution story (i.e., a narrative that articulates how research [or, in this context, a research portfolio] contributed to a claimed result and describes the nature of this contribution) (Mayne, 2001). A contribution story acknowledges the existence of other factors that might have played a role in achieving the claimed results and collects evidence for each component of the storyline. This, in turn, enables an interrogation of the story in light of this evidence.

The use of qualitative approaches (ideally supported by quantitative metrics) as a key principle is also explicitly highlighted in existing literature, including the international Leiden Manifesto for Research Metrics developed under the leadership of scholars from the Georgia Institute of Technology and Leiden University (Hicks et al., 2015).[1] The value of qualitative approaches was emphasized by multiple interviewees, who viewed narratives as an important tool for taking stock of ongoing research activities and their results. Although narratives might be particularly important for applied and translational research activities, because it is easier to talk about how people can use the outputs of such research, even such organizations as NSF that focus on basic research utilize narratives because they help to explain the value of research to laypeople. To illustrate, NASA, which supports a range of research and the application of research in new technologies, produces a book and associated webpage annually to discuss its commercialization of technologies developed for space missions that get additional uses in completely unrelated areas (NASA, undated). Another example is the impact stories presented in a review report on the second decade of the National Occupational Research Agenda (Felknor, Williams, and Soderholm, 2017).

The disadvantages of the narrative approach are twofold and relate to practicality. First, the collection of necessary evidence and the subsequent production of the narrative assessment can be particularly costly and place substantial burden on researchers or research administrators.[2] Second, because narrative assessments aim to uncover the unique contribution of underlying research, they are not well suited to comparisons across various contexts. On the other hand, they can be useful for comparisons within in a given area, as demonstrated by their use in the UK's Research Excellence Framework (REF), an impact-assessment process,[3] although such comparisons might not be available in response to quick-turnaround requests for information. Narrative assessments generally do not result in summary measures.

---

[1] See also the dedicated webpage: Leiden Manifesto for Research Metrics, undated.

[2] This is a heuristic assessment of the level of effort needed to construct portfolio-level metrics of similar levels of complexity. We revisit this point in Chapter Four.

[3] We also discuss the REF in greater detail in the section on time lags below.

General (e.g., Population-Level) Metrics

A third approach to assessing the performance of research portfolios involves general, typically population-level, metrics that are linked in some way to the objectives of the underlying research. For instance, a research portfolio addressing a particular disease might aim to improve treatment, thereby reducing the disease's burden among the target population. Trends in the disease burden might be used as a metric to assess the research portfolio. Such metrics are typically very high level by nature.

This approach has several notable advantages. First, like narratives, general metrics are likely to be understandable by wide audiences and will capture issues that are of importance for the underlying research. The metric can also be used for longitudinal trend analyses. As such, it can be particularly attractive to decision makers. In addition, relevant data are likely to be available (e.g., in existing public health databases) and might not require data-collection efforts on the part of researchers. In light of these advantages, general population-level metrics might be of special interest in the context of DoD research related to health and health care, which aims to improve the outcomes for a clearly definable (albeit heterogeneous) population served by a closed health care system.

In the absence of corroborating evidence, the use of such metrics is subject to substantial attribution issues (see the discussion in the concluding section of this chapter) because these metrics cannot shed light on the extent to which the observed changes are attributable to the research activities under assessment. Nor can they illuminate what the trends in the observed metric would have looked like in the absence of the research. The use of a logic-model approach, however, can help provide evidence of the pathways by which the components of a portfolio address its ultimate objectives.

Common Challenges

The application of research-portfolio metrics is not immune to common challenges associated with research evaluation. In this section, we discuss the concerns we found most notable in assessing research portfolios. They include (1) heterogeneity of portfolios, (2) quality and availability of data, (3) research time lags, and (4) challenges associated with attribution or contribution.

Heterogeneity of Portfolios

Not all research portfolios are well suited to aggregate metrics. This is typically because an agency can conduct or fund research in a variety of areas with varying objectives that do not support meaningful comparisons. Consequently, with respect to assessing the results of completed research, some organizations (e.g., FDA, NIST, NIOSH) conduct research evaluations primarily at the program level. These research programs still meet the definition of *portfolio* because each brings together a body of research consisting of multiple individual projects or studies.

Evaluation of heterogeneous portfolios as a whole needs to adhere to the following principles:

- Upstream metrics (inputs, processes, and outputs) are much better suited than downstream metrics to cope with portfolio heterogeneity. Examples of metrics applicable in

this context include various breakdowns of funding (e.g., by topic area, by type of study, by stage of research), breakdowns of activities (e.g., number of studies per topic area), and output aggregates (e.g., number of patents, publications).
- Downstream measurements might require the utilization of high-level metrics (e.g., as seen in GPRA reporting) or the development of narratives of success stories, both of which are subject to limitations discussed earlier in this section.[4]

Quality and Availability of Data

As discussed above, portfolio-level assessments rely on data that are frequently collected at the level of individual projects or studies. The availability of such information is an important prerequisite for the ability to construct portfolio-level metrics. Data feeding upstream metrics (inputs, processes, and outputs) are frequently provided by existing research administrative databases and repositories of information on research projects, which can be used to generate assessment reports on the research portfolio as a whole (NIH's RePORTER [NIH RePORTER, 2018] and Federal RePORTER [STAR METRICS, 2018] are examples of repositories that enable a range of assessment products). Other non–public-facing systems that serve as repositories of relevant data are NIH's IMPAC II database and the Electronic Grants System in use by the CDMRP. Gore and her colleagues explored the feasibility of establishing similar data repositories for DoD psychological health (PH) research (Gore et al., unpublished).

Data used to populate agencies' administrative databases and data repositories are frequently collected for researcher-provided progress and final reports. Seven testimonies pointed to annual, midyear, and closeout study reports, as well as in-progress reviews and annual portfolio reviews and analyses, as important or primary sources of data. A notable initiative in this context is the Research Performance Progress Report, developed by NSF and adopted (with modifications) by at least 12 agencies for the purposes of annual or other interim research progress reporting (NSF, undated [c]). The report format consists of a series of mandatory, as well as optional, modules. In addition, agencies have the possibility to develop modules specifically tailored to their respective contexts (subject to OMB approval), although, in the interest of uniformity, they are requested to minimize any supplements (NSF, undated [b]).

With respect to downstream measures on outcomes and impacts, collecting data typically requires gathering information from external sources. Examples of efforts in this area that interviewees mentioned include reviews of publicly available data repositories (e.g., publication and patent databases), as well as requests for data searches from relevant external organizations (e.g., adopters or beneficiaries of new tools or guidelines). Furthermore, eight interviewees mentioned the use of site visits, interviews, and surveys as data-collection tools, although two interviewees expressed skepticism about the utility of surveys. Three interviewees also highlighted the possibility of commissioning external evaluations, which can be particularly attractive for organizations with limited in-house evaluation capacity. All of these steps can be time-consuming and costly. Correspondingly, as most interviewees pointed out, agencies are faced with assessing what data merit collection given available resources, a decision that, in turn, shapes the breadth and depth of research-portfolio assessments. One interviewee added that a likely outcome of this consideration is that, because organizations cannot measure everything

[4] This is not to say that some aggregations of downstream measures are not possible. For instance, impact arrays (i.e., graphical representations of impacts from organizations' research portfolios), utilized as part of RAND's ImpactFinder tool, are one example of efforts in this direction (Grant, 2012).

they want, they settle for less desirable, but easier-to-obtain, metrics. A small number of stakeholders also pointed out that external users of the research products are likely to face similar resource constraints on data collection. For instance, one interviewee stressed that clinical providers do not have the resources or formal mechanisms in place to track all desirable metrics, limiting what could be fed back to the research organization.

CEP highlighted several data-related constraints related to evaluations and evidence-building activities. It noted that agencies' existing administrative processes are not always well designed to support evidence building, and it stressed that some agencies face difficulties in identifying funding for evidence-building activities, including necessary data collection. In some contexts, responsibilities for evidence-based activities and data collection can be distributed across existing departments, a distribution that can give rise to coordination challenges (CEP, 2017). The commission identified the HHS Data Council, established to facilitate intra-HHS coordination on data collection and analysis, as an example of successful data resource coordination efforts (HHS, undated).

One interviewee suggested that, within the DoD context, downstream data collection can be particularly problematic for research leading to the development of knowledge products. The interviewee opined that, although, for new materiel, one entity might be responsible for its acquisition, storage, as well as sustainability, which could be a source of data on the uptake of the new product, there is typically no such arrangement in place for knowledge products. In addition, this challenge can be compounded by the fact that entities developing knowledge products do not see looking beyond the development of the product itself as part of their mission and thus might not have formal outcome and impact targets.

To address the challenge of collecting necessary data, several interviewees suggested, it might work best to have someone "riding along" at an administrative level to document and record. One example of such an arrangement is the Hearing CoE, which has research coordinators assigned to various military treatment facilities who are able to monitor the execution of relevant research. The Defense and Veterans Brain Injury Center (DVBIC) is also able to use similar networks for its data collection. The deployment of staff dedicated to data collection, of course, requires resources, as well as training to ensure the knowledgeability of the collecting personnel.

Last, a small number of interviewees also opined that, in some instances, the main challenge to assessing research performance is not data availability but data's quality or amenability to analysis and measurement. For instance, interviewees mentioned new and emerging approaches to mining and analyzing textual and other data, which suggest possibilities for useful analyses and inferences with minimal reporting burden for researchers.[5]

Research Time Lags

A well documented challenge in research evaluation is research time lags (i.e., the fact that it can take a relatively long time before any impacts resulting from research materialize). It can take even longer for impacts that depend on enabling circumstances (e.g., achievement of different kinds of readiness, advances in complementary technology). Furthermore, the length of this time lag can vary considerably across contexts and is difficult to measure. For instance, in

---

[5] See, for example, the work of ÜberResearch and its dimension database (ÜberResearch, undated), tools developed by altmetrics (altmetrics, undated), and F1000Prime, which combines peer review, data mining, and crowdsourcing (F1000Prime, undated).

a series of case studies, Guthrie and her colleagues examined the time lags and attribution in the translation of cancer research (i.e., the time elapsed between the funding of research and health gains) and found substantial variation in the time frames observed, as well as a range of factors contributing to the elapsed time (Guthrie, Pollitt, et al., 2014). Similarly, a review by Morris, Wooding, and Grant found substantial limitations in the existing evidence base on research time lags (Morris, Wooding, and Grant, 2011). They did note that some studies arrived at an estimate of 17 years (Balas and Boren, 2000; Grant, Green, and Mason, 2003; Wratschko, 2009), although these estimates utilized different approaches and end points. Of course, the time lags for organizations engaged in applied or translational research (e.g., along the lines of activities of the CoEs) can be expected to be shorter because the clock starts running later in the process. However, this distinction does not remove the fundamental challenge of relevant data not being available at the conclusion of research activities and their associated funding.

A practical implication of this challenge is that the time frame for the collection of data on research outcomes and impacts can be very long and extend until multiple years after research completion (as schematically captured in Table 2.1). This gives rise to challenges related to such issues as allocating the responsibility for and providing resources to such data-collection work. As discussed above, some agencies make a point of searching literature and even patents for pointers to their sponsorship of underlying research. Ongoing developments in new information technology tools (discussed in Chapter Four) can be expected to continue making such endeavors more accessible to research organizations. Other agencies have reported efforts to reach out periodically to former grant holders to inquire about any results arising from the research and (where applicable) its commercialization.[6] Of course, agencies' ability and willingness to reach out to past performers vary.[7]

Another example of a research funder's attempt to address this challenge is to make a small tranche of funding contingent on the production of an impact report, submitted one year after the completion of the research.[8] Although this time frame is still very likely not sufficient for major impacts to materialize, it nevertheless provides an opportunity to collect data on short-term outcomes and impacts, as well as additional information that might shed light on the likelihood of achieving medium- and long-term impacts.

Table 2.1
Time Frame for Data Collection

Research Stage	Input	Activity	Output	Outcome	Impact
During	x	x	x		
After			x	x	x

[6] One example volunteered by an interviewee was to establish a phone bank and call every previous holder of a Small Business Innovation Research contract with the agency two years after its completion to ask about any results and whether the agency can be credited, at least partly, with their achievement.

[7] Context and culture factor in: In the UK, for example, as a result of the Research Councils UK Research Outcomes Harmonisation Project, a condition of government funding through the national Research Councils is impact reporting via the Researchfish site, which is also used by research funders elsewhere in Europe and in the United States.

[8] This is an approach taken by Growth and Labour Markets in Low Income Countries (labor-market research for development) (Growth and Labour Markets in Low Income Countries, 2017).

Yet another example is the REF, whereby academic researchers are required to demonstrate the impact of their research (the definition of *portfolio* being de facto left up to researchers and their institutions because they retain editorial control over what body of research they will submit for assessment) (REF, undated [a]). This process places the burden of evidence collection on the researchers and might incentivize them to have continuous data-collection systems in place (see, e.g., Jones, Manville, and Chataway, 2017, and Manville et al., 2015). The REF's design also explicitly acknowledges time lags in its eligibility criteria: In the latest iteration in 2014, submissions needed to describe impacts that occurred between 2008 and 2013 and had to be based on research undertaken between 1993 and 2013 (REF, 2012).[9]

Challenge of Attribution

With respect to downstream measures, such as outcomes and impacts, attribution represents a critical challenge in that it can be very difficult to establish a link between any observed or claimed outcomes and impacts and the underlying research. In this context, the term *attribution* is frequently used interchangeably with *contribution*, although a clear distinction can be drawn (Morgan Jones and Grant, 2013). *Attribution* typically relates to the extent to which underlying research led to the production of outputs or outcomes (or the proportion of that underlying research that did so). By contrast, *contribution* focuses on the ability to establish the claim that outputs and outcomes have resulted from the underlying research, irrespective of the relative volume of such contribution. Focusing on contribution can be particularly important if the objectives of research evaluation are linked to advocacy or accountability purposes (CAHS, 2009).

Here is an illustration of the challenges in this area: Some organizations might use trends in disease burden as a metric of the impact of their research. In this context, it is very difficult to demonstrate the research's unique contribution to the observed trends, which are likely a product of numerous simultaneous factors (or multiple strands of research). Similarly, a common outcome metric for research portfolios is the level of uptake of new tools or clinical procedures, which factors external to the underlying research can affect. Some agencies might not even view supporting the uptake of their research products to be part of their mission or objectives.

The majority of stakeholders who commented on impact measurements by their agencies, although they considered impact metrics very desirable, stressed the difficulty of measuring the uptake of research and its impact. One contributing factor that several interviewees highlighted was the fact that any potential outcomes and impacts that might materialize will be a few steps removed from the originating research organization. This leads not only to data-collection issues but also to challenges surrounding the tracking of any causal links between individual steps that might have ultimately led to the outcomes and impacts of interest, not least because, as one interviewee emphasized, data on such metrics as patient outcomes are invariably noisy. Thus, as one interviewee summarized, "impact on health is the most important but hardest thing to measure."

---

[9] For a small number of research fields, the period of eligibility for the underlying research started in 1988. The eligibility criteria also acknowledged the possibility that some impact might occur before the results of the research are formally published. For that reason, the eligibility period for research outputs ended six months after the eligibility period for impacts (December 2013 and June 2013, respectively). The upcoming REF 2021 calls for assessing impacts between 2013 and 2020 associated with research between 2000 and 2020 (REF, 2019 [b]).

One solution to this challenge is to develop a narrative accompanying the impact metrics that describes the research portfolio's unique contribution (see, e.g., Mayne, 2001). In this regard, considering the counterfactual (i.e., what would have happened or how the situation would have looked in the absence of the research portfolio and its products) might be useful. Recognizing the need for narratives to isolate the attribution or contribution of underlying research, some of the approaches to research evaluation with the most creativity and innovation have been international. From that context, it is worth noting that (1) complementary use of quantitative and qualitative metrics is one of the key principles in the Leiden manifesto (Hicks et al., 2015) and (2) the use of narrative case studies demonstrating the impact of underlying bodies of research is the basis of the UK's REF.[10] In the U.S. context, a range of agencies and organizations, including NASA, NIH, NSF, and NIOSH, employ outcome and impact narratives.

---

[10] See Appendix A for other insights from international comparisons.

CHAPTER THREE

Overview of Identified Metrics

Summary of Main Points

The most frequent type of metrics identified in this review pertained to outputs, followed by those related to outcomes.

The large number of metrics in the output category is not surprising given the continued emphasis on publishing as a measure of research productivity and the variety of output types that could be produced.

Non-DoD organizations appear to place greater emphasis than their DoD counterparts that we examined do on outcome and impact metrics.

During our review, we did not observe any notable differences in the use of metrics by organizations that focus exclusively on medical research and use by those having medical research as just one of their focal areas.

Assessed against our framework's three criteria, upstream metrics (inputs, processes, and outputs) tend to be rated in a similar fashion:

- They generally do not require much expert judgment.
- Data burden associated with their collection is relatively low.
- They do not tend to give rise to attribution issues.

By contrast, the assessment of downstream metrics (outcomes and impacts) fully demonstrates the trade-offs associated with their use.

This chapter presents the metrics identified in the course of this study. Several prefatory observations are worth noting.

First, as discussed in the "Methodology" section in Chapter One, our analysis categorized individual metrics by (1) stage of measurement, (2) broad category of metric, and (3) concrete metric type. To illustrate, the metric benefit-to-cost ratio is categorized as (1) an impact metric (at the stage of measurement), (2) a cost-effectiveness or economic-return metric (a broad category), or (3) a return-on-investment (ROI) metric (a concrete metric type).

Relevant agencies operationalize each type of metric to suit their needs and contexts. In some instances (e.g., the number of publications), the concrete form of the individual metric is identical or very similar across all user agencies. In other instances, there are notable differences in how agencies implement a particular type of metric. For instance, citation-based indicators can take several forms, ranging from a count of citations to Relative Citation Ratios (RCRs).

Second, an agency can use more than one form of a particular metric type. For that reason, an indication of a count of metrics in a given metric type does not always equal the number of agencies using this type of metric. As such, the frequency numbers presented in the figures in this chapter capture the number of agency–metric pairs within each metric type, rather than exclusively a number of agencies utilizing a given type of metric.

Third, it is important to keep in mind that, during our review, we could positively identify only instances of individual metrics being used or proposed for use by individual agencies. With the information collected, it is not possible to determine that a given metric is *not* in use by a given agency, and presented findings should not be interpreted as such. In other words, the actual use of metrics could be higher than it appears based on the reported frequencies.

Fourth, it needs to be recognized that the classification of and categorization of individual metrics into types and categories is, to some extent, a product of our judgment, as well as a function of data availability, presentation, and granularity. To illustrate the latter point, where some documents might distinguish multiple individual metrics (e.g., number of publications and number of peer-reviewed publications separately), other sources might not make this distinction explicit even though they make it in practice.

A similar point could be made about the categorization of individual metrics across the five stages of measurement captured in the conceptual logic model. In numerous instances, a case could be made for more than one form of categorization, with individual metrics potentially straddling the boundaries across adjacent categories. This is particularly the case with respect to outputs, outcomes, and impacts. In our categorization, we took note of how individual agencies viewed and described metrics they used, although judgments were necessary for the purposes of cross-agency aggregation. Perhaps most notably, in regard to the differentiation between outputs and outcomes, the research team took the view that outputs rest firmly within the control of the researchers, while outcomes do not have to be. As such, such metrics as the number of publications and other products are discussed under outputs, while the uptake of these products by third parties (e.g., journal citations, media mentions) are classified as outcomes.

Fifth, we reiterate a crucial qualification that the frequency of the use of a particular metric is not necessarily an indicator of the metric's quality and appropriateness (irrespective of whether the metric is contextually relevant for its user agency). The fact that a notable number of agencies report using a given metric does not automatically mean that it is a good one. For instance, despite the well documented limitations of a journal impact factor as a metric (discussed in greater detail below), some research organizations continue to use it.

For these reasons, we urge caution in interpreting the quantifications associated with the presentation of metrics. In the opinion of the research team, the utility of metric frequencies lies in providing an indication of the breadth of individual metric types, as well as a general sense of the relative frequency of their use. From our point of view, given the caveats mentioned above, detailed statistical analysis of available numeric indicators, as well as analysis pertaining to individual agencies, is neither advisable nor appropriate.

Overall Distribution of Metrics

We begin our discussion of identified metrics with a general overview of their distribution across a variety of criteria. Figure 3.1 shows the distribution of metrics per stage of measurement. The largest number of identified metrics pertains to outcomes, followed by those pertaining to outputs.

Figure 3.2 shows the distribution of metrics by stage of measurement, broken down by whether the metric in question has been used or has been suggested to be used. The breakdown shows that even using only metrics that have definitely been used does not substantially alter their distribution across stages of measurement.

Figure 3.3 shows the distribution of metrics by type of organization (DoD, non-DoD U.S. government, or organizations outside of the federal government). Among the metrics

Figure 3.1
Distribution of Metrics, by Stage of Measurement

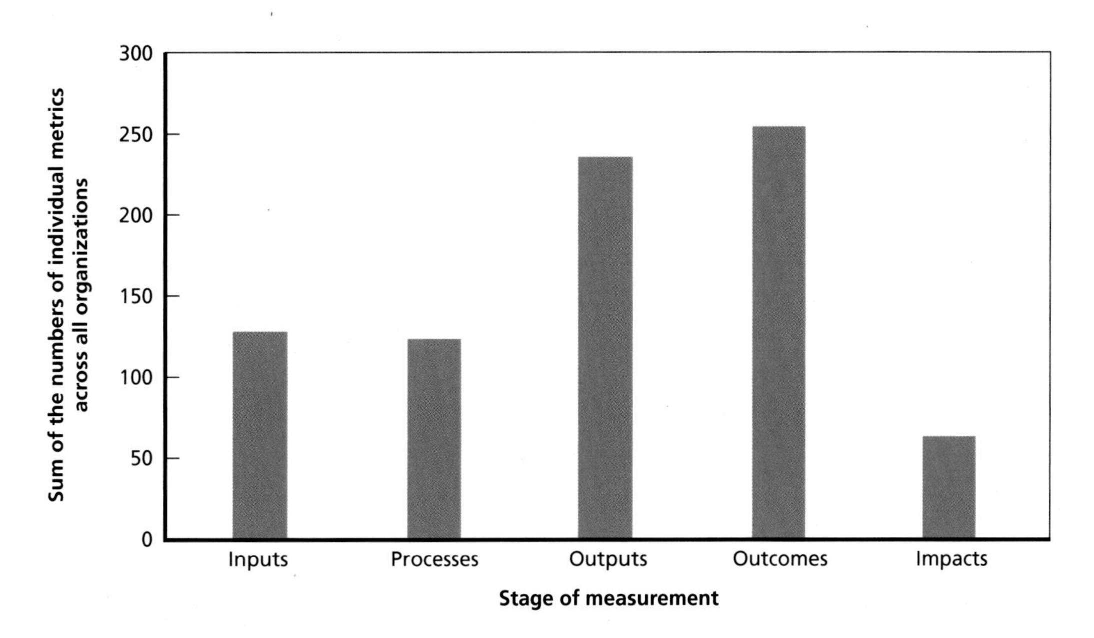

Figure 3.2
Distribution of Metrics, by Type of Metric

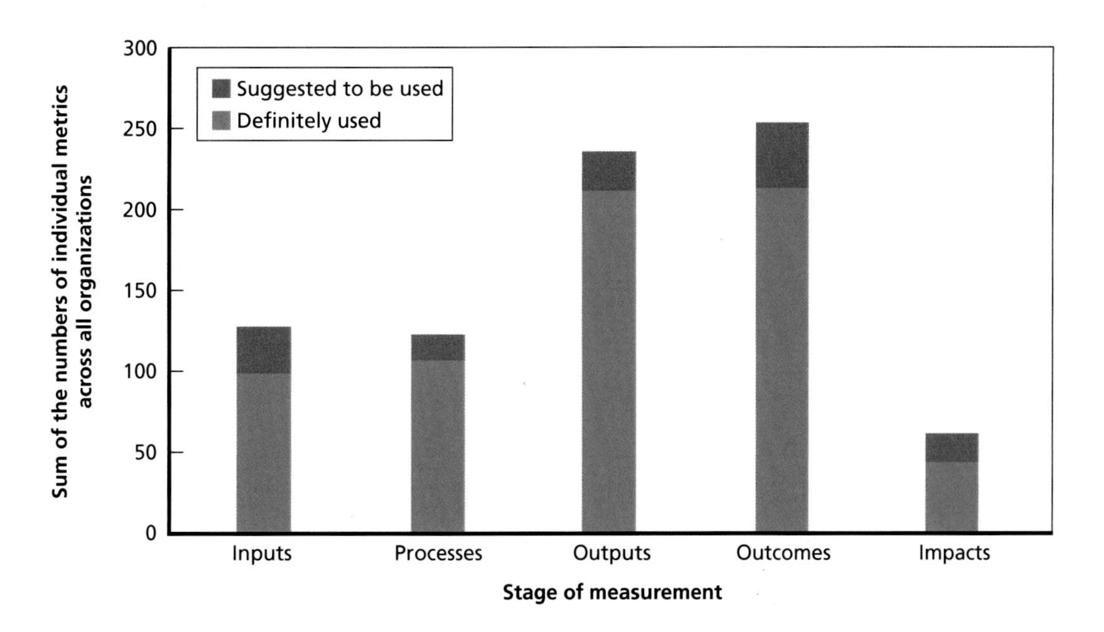

identified as part of this review, output metrics were the largest group for DoD and non-DoD U.S. government organizations; for private, non–U.S. government organizations, the largest number of identified metrics pertained to outcomes. Furthermore, although the number of identified metrics associated with DoD agencies that we examined was lower than that for the other two types of organizations at every stage of measurement, the difference was notably larger for outcome and impact metrics than for measures further upstream. Testimony from

Figure 3.3
Distribution of Metrics, by Type of Organization

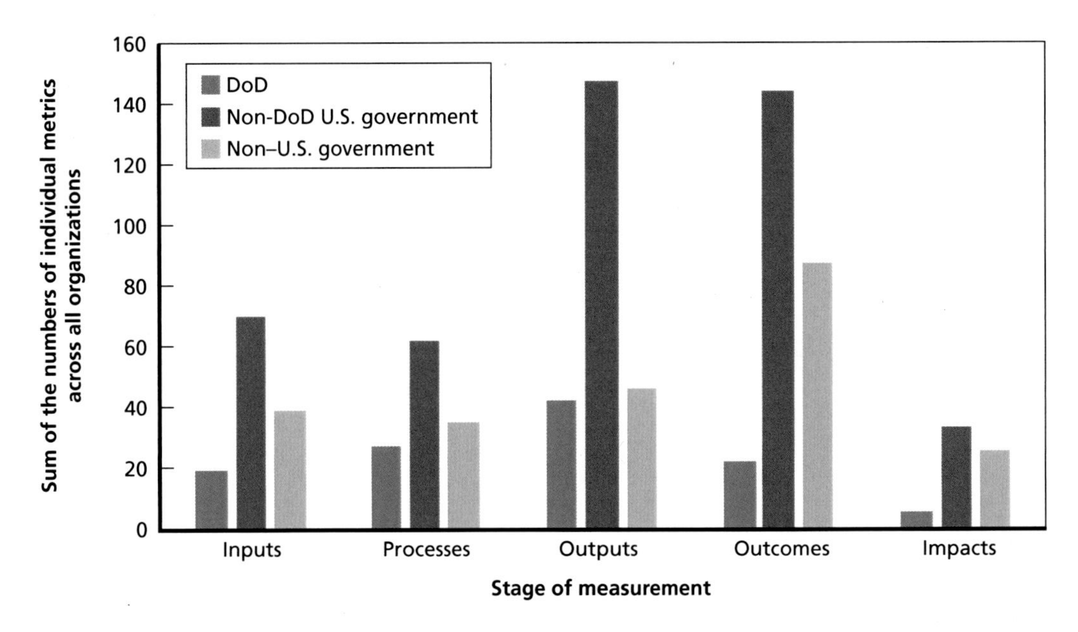

stakeholders also corroborated our observation that non-DoD agencies focused more than the DoD components we examined on measuring outcomes and impacts than on measures further upstream.

Figure 3.4 presents the distribution of metrics by agency focus (i.e., it disaggregates the count of metrics at each stage of measurement by whether the agency in question focuses exclusively on medical research or whether its research activities involve research in other fields,

Figure 3.4
Distribution of Metrics, by Agency Focus

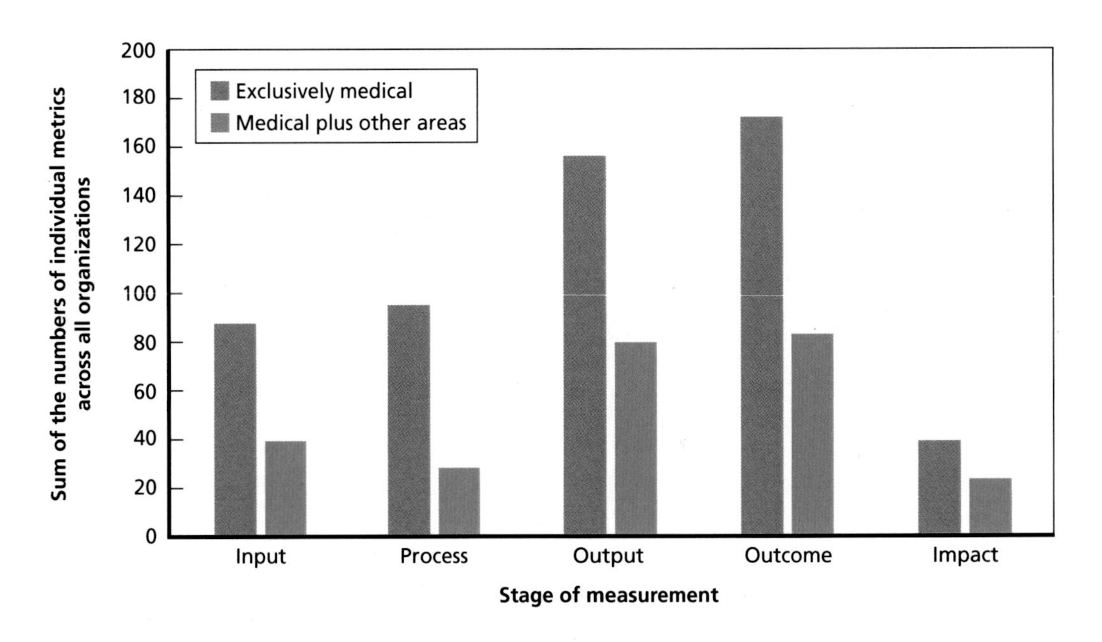

such as science and technology). This disaggregation did not reveal any notable differences between the two categories of agencies.

The rest of this chapter discusses metrics identified for each stage of measurement. Additional extractions from our database of metrics can be found in Appendix C.

Input Metrics

Metric Types

Figure 3.5 presents the relative frequency of individual input metric types. The metric type with the greatest variety is probably indicators of investigator quality and organizational prestige, for which agencies might use a variety of concrete individual metrics. Under indicators of investigator quality, the individual metrics identified in this review include percentage or number of principal investigators (PIs) with professional recognition as captured by various affiliations (such as membership of journal editorial boards, peer review panels, federal advisory councils, editorships of high-profile journals, relevant boards and committees, academy membership, or leadership roles in professional societies).

Under indicators of organizational quality, possible individual metrics include retention rates of key staff and the number of (out-of-state and international) applications per job opening (on the assumption that high-quality research institutions do better at attracting and retaining staff). Other identified metrics under this type include the track record of new hires and staff career progression.

Assessment

Table 3.1 presents an assessment of input metrics against the three criteria included in our assessment framework, as discussed in the "Assessment of Metric Data" section in Chapter One: the degree to which a given type of metric incorporates expert judgment, the relative

Figure 3.5
Identified Types of Input Metrics

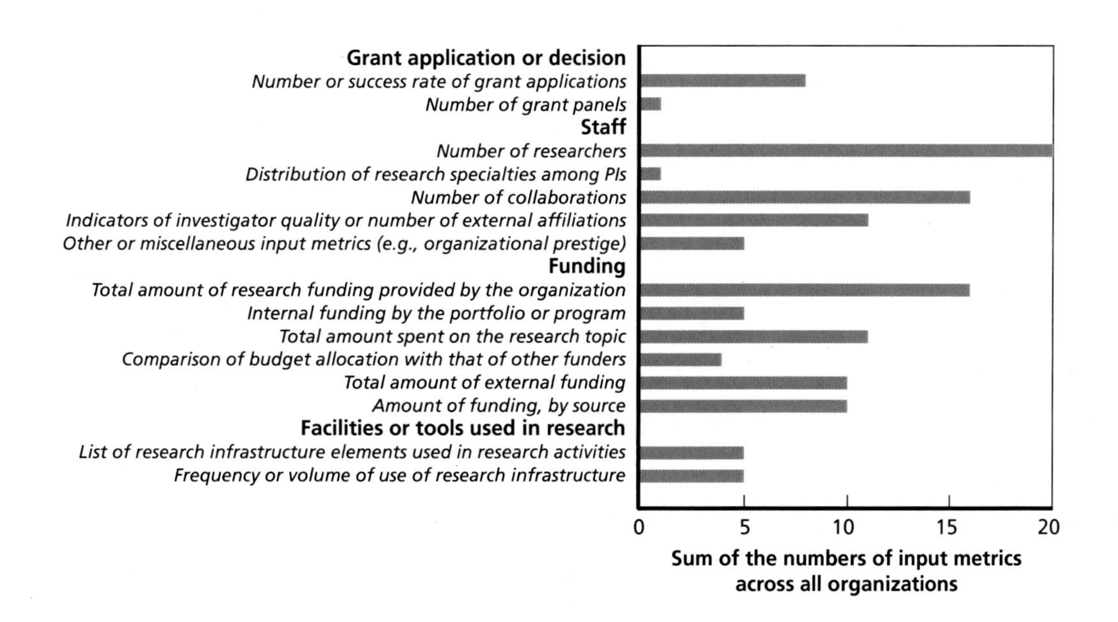

Table 3.1
Assessment of Input Metrics

Metric	Reliance on Expert Judgment	Data Burden	Attribution Issues
Grant application or decision			
Number or success rate of grant applications	Low	Low	Low
Number of grant panels	Low	Low	Low
Staff			
Number of researchers	Low	Low	Low
Distribution of research specialties among PIs	Low	Low	Low
Number of collaborations	Low	Low	Low
Indicators of investigator quality or number of external affiliations	Medium	Low	Medium
Other or miscellaneous input metrics (e.g., organizational prestige)	Medium	Low	Medium
Funding			
Total amount of research funding provided by the organization	Low	Low	Low
Internal funding by the portfolio or program	Low	Low	Low
Total amount spent on the research topic	Low	Low	Low
Comparison of budget allocation with that of other funders	Medium	Medium	Medium
Total amount of external funding	Low	Low	Low
Amount of funding, by source	Low	Low	Low
Facilities or tools used in research			
List of research infrastructure elements used in research activities	Low	Low	Low
Frequency or volume of use of research infrastructure	Low	Medium	Low

burden associated with the collection of data necessary to inform the type of metric, and the degree to which a given type of metric gives rise to attribution issues. The vast majority of input metric types score low on all three criteria. This is perhaps not surprising in that input metrics can generally be informed by existing administrative data that do not require much interpretation, tend to be readily available, and are clearly related to the underlying research.

One point to note about this set of metrics is that, for some, there might not be an obviously desirable set of values these should take or a direction of change one would like to observe over time. Put differently, it is not always possible to say that higher input values are preferable to lower values. For example, the appropriateness of the total amount spent on the research topic depends on the scope of the research topic and how the topic is defined. It is difficult to compare seemingly similar topics if the scopes are widely different. For example, narrow scope might dictate less funding, while a potentially transformative project on the same topic might require more funding, and a narrowly scoped, relatively inexpensive project can be valuable or at least cost-effective. Similarly, having a large number of researchers might not always be preferable and might indicate low cost-effectiveness; a low number might be sufficient (preferable) in some cases.

Process Metrics

Metric Types

Figure 3.6 presents the relative frequency of individual process metric types. The majority of metric types are self-explanatory and revolve around enumerating and aggregating various types of activities and their subtypes. Other metric types aggregate various descriptors of research activities, such as details on the populations covered, topics investigated, and study types employed (e.g., clinical trials, evidence reviews). Several agencies explicitly compare the composition of their research portfolios to their organization's stated aims, although it can probably be assumed that other agencies examining the composition of their portfolios (e.g., through number of projects by research topic) are able to do so as well. This comparison can rely on expert assessments but can also take the form of contrasting the composition of the portfolio with the objective criteria set out in programmatic and strategic documents.

Options to capture administrative efficiency and burden include measurements of the burden in absolute terms (e.g., hours or days spent fulfilling administrative obligations), as well as those in relative terms (e.g., in terms of percentage of spending). Multiple options also exist with respect to measuring research progress toward milestones, which are typically determined at the outset of each study (e.g., IRB approval, recruitment, deliverables). One possibility is to capture the number or percentage of projects that have met their milestones. Another option is to take stock of projects that have had notable reportable events or issues that might affect their

Figure 3.6
Identified Types of Process Metrics

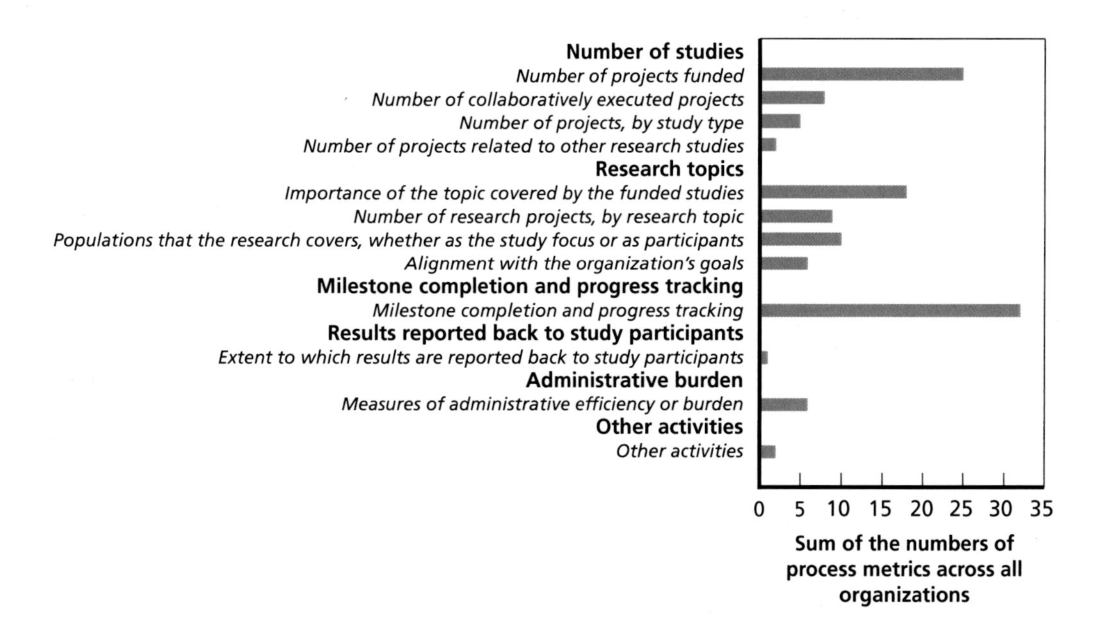

ability to meet milestones in the future. Requests for additional time through for- or no-cost extensions and other changes to future plans can also be used in these assessments.

Assessment

As with input metrics, the majority of process metric types score low on the three criteria used in our assessment framework (Table 3.2). The exceptions are metric types that will likely require a somewhat higher degree of expert judgment and consideration. These include an assessment of relations and connections between individual studies, as well as assessments of the relative importance of topics covered by ongoing research projects and their alignment with organizational goals. Some expert judgment might also be required in the interpretation of certain metrics related to milestone completion and progress tracking. For instance, an elevated number of extensions cannot be automatically understood as a negative development,

Table 3.2
Assessment of Process Metrics

Metric	Reliance on Expert Judgment	Data Burden	Attribution Issues
Number of studies			
Number of projects funded	Low	Low	Low
Number of collaboratively executed projects	Low	Low	Low
Number of projects, by study type	Low	Low	Low
Number of projects related to other research studies	Medium	Low	Low
Research topics			
Importance of the topic covered by the funded studies	Medium	Low	Low
Number of research projects, by research topic	Low	Low	Low
Populations that the research covers, whether as the study focus or as participants	Low	Low	Low
Alignment with the organization's goals	Medium	Low	Low
Milestone completion and progress tracking			
Milestone completion and progress tracking	Low	Low	Low
Results reported back to study participants			
Extent to which results are reported back to study participants	Low	Low	Low
Administrative burden			
Measures of administrative efficiency or burden	Low	Low	Low
Other activities			
Other activities	Low	Low	Low

and it needs to be considered in the context in which these developments occurred. Similarly, indicators of milestone completion can be skewed if the milestones were set inappropriately in the first place.

Output Metrics

Metric Types

Figure 3.7 presents the first group of output metric types. The majority of metric types in this group are self-explanatory and revolve around the counts of various types of products. These range from various types of publications to new devices and tools to patents. Of course, metrics based on a particular type of output are meaningful only for activities that can reasonably be expected to result in such outputs. For instance, measuring the production of journal publications might not be appropriate for research activities other than the kinds of basic and applied research that typically get published in journals.

Two metric types merit further elaboration. First, various individual metrics fall under "publication quality." Most frequently, an assessment is made by examining the impact factor of the journals in which relevant publications appear. The aggregate portfolio-level metric could then take the form of an average impact factor across the portfolio, as well as a number of articles published in journals above a certain impact factor. However, as pointed out by an interviewee, as well as existing literature, using journal impact factors has its well documented shortcomings (see, e.g., Lariviere et al., 2016; Callaway, 2016). Another possibility for arriving at this metric is an internal assessment of what publications in the portfolio are of high quality.

Second, the use of patents as an output metric can be beset with interpretation and methodological challenges. The assumption that patents can be seen as a proxy measure for research impact is complicated by the fact that a substantial share of patents do not result in any subsequent research or commercial activities. Relatedly, large numbers of patents are filed

Figure 3.7
Identified Types of Output Metrics

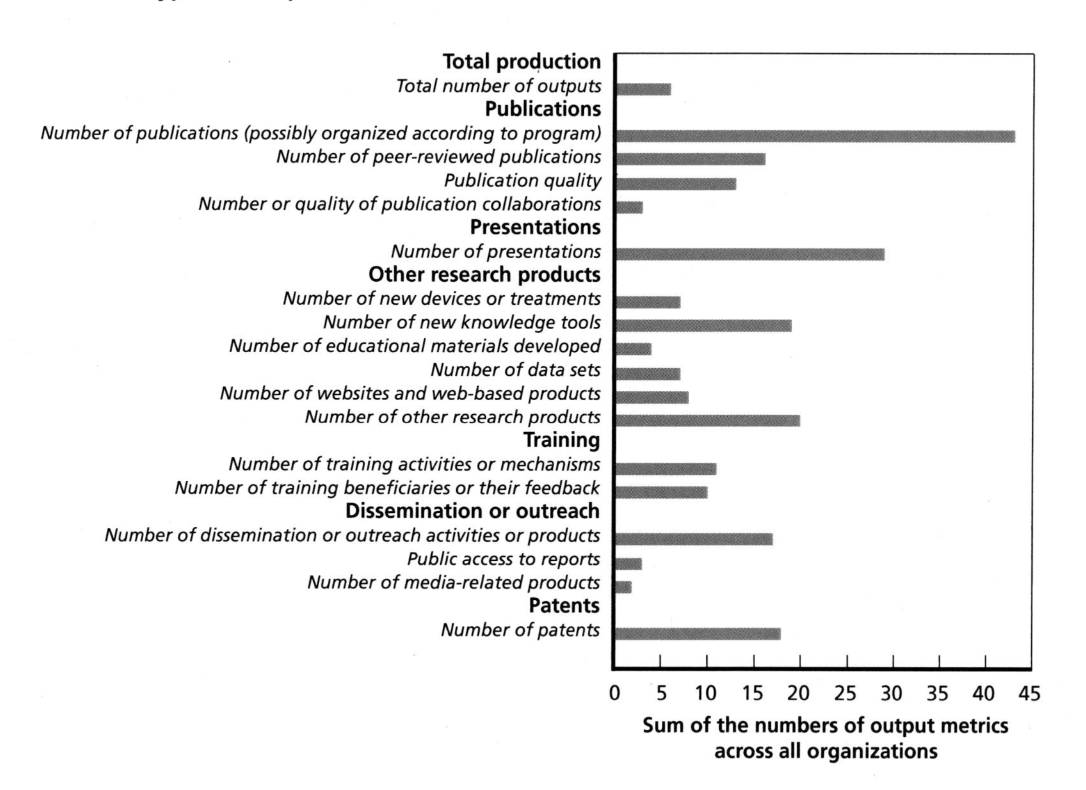

for defensive purposes without any immediate follow-on plans (see, e.g., Sampat and Williams, 2017; Ledford, 2013).

The remainder of identified output metric types revolve around training-related products, as well as dissemination efforts and results building on the underlying research. With respect to the results of dissemination activities, some agencies keep track of the extent to which their research products are publicly available (e.g., on the organizational website or through open-access repositories). Some agencies track the extent to which research products have reached relevant audiences (e.g., by collecting data on the number of clinicians and health professionals their dissemination efforts reach).

Assessment

The assessment of output metrics against the three criteria is presented in Table 3.3. As with input and process metric types, the prevalent rating across all three criteria is low, reflecting the fact that the metrics generally take the form of an aggregate numeric indicator drawing on data that are generally easily available. However, we reiterate that some degree of expert judgment will still be involved for at least some of these metric types, particularly with respect to their design, definitions, and conceptualizations.

Outcome Metrics

Metric Types

Figure 3.8 presents the variety of types of metrics identified to capture research outcomes. The first set of metric types revolves around an examination of how much other researchers and other groups or entities acknowledge and reference the research outputs. Multiple options exist under the metric type of citation-based indicators, although these can give rise to methodological and interpretation challenges. As one interviewee opined, citations are a useful proxy for publication influence, both in terms of positive influence (informing further publications) and negative influence (when cited to disagree). Many organizations keep track of the absolute numbers of instances relevant publications were cited,[1] as well as the proportion of highly cited publications (e.g., those in the top 5 percent of the field). Another possible metric is to calculate an institutional h-index, i.e., an index that expresses the number of publications that have been cited at least h times (Hirsch, 2005). As such, the h-index captures both volume of production and publication impact. However, the h-index is subject to a series of limitations as well. First, the value of the h-index can grow over time even in the absence of new publications. Second, its value can depend on the database used for its calculation (e.g., Google Scholar versus Web of Science). Third, and perhaps most importantly, it is field dependent (i.e., researchers in certain fields are likely to have higher h-indexes (Hicks et al., 2015). To address this challenge, it is necessary to field-normalize available citation data. A recent tool in this regard, developed and deployed by NIH researchers, is the RCR, which makes use of papers' co-citation networks (Hutchins et al., 2016).

Metrics pertaining to nontraditional methods of engaging with research products (e.g., through new communication platforms) are in line with a recent increased focus on novel methods of assessing the reach of research products, such as webometrics and altmetrics (Martín-

---

1 When aggregated, these data would ideally be field and year normalized.

Table 3.3
Assessment of Output Metrics

Metric	Reliance on Expert Judgment	Data Burden	Attribution Issues
Total production			
Total number of outputs	Low	Low	Low
Publications			
Number of publications (possibly organized according to program)	Low	Low	Low
Number of peer-reviewed publications	Low	Low	Low
Publication quality	Low	Low	Low
Number or quality of publication collaborations	Low	Low	Low
Presentations			
Number of presentations	Low	Low	Low
Other research products			
Number of new devices or treatments	Low	Low	Low
Number of new knowledge tools	Low	Low	Low
Number of educational materials developed	Low	Low	Low
Number of data sets	Low	Low	Low
Number of websites and web-based products	Low	Low	Low
Number of other research products	Low	Low	Low
Training			
Number of training activities or mechanisms	Low	Low	Low
Number of training beneficiaries or their feedback	Low	Medium	Low
Dissemination or outreach			
Number of dissemination or outreach activities or products	Low	Low	Low
Public access to reports	Low	Low	Low
Number of media-related products (e.g., press releases)	Low	Low	Low
Patents			
Number of patents	Low	Low	Low

Martín et al., 2016). Altmetrics (or alternative metrics) are research indicators based on activity on social media platforms. These platforms can take many forms, including social networks, blogs, videos, and ratings and recommendations on social sites. However, although altmetrics have enjoyed popularity and interest, several interviewees, as well as existing literature, raised a series of concerns regarding their use for the purposes of research evaluation. These concerns include (1) various altmetric aggregators' use of digital object identifiers, which leaves out certain types of research products, as well as discussions of those products in subsequent blogs

Figure 3.8
Identified Types of Outcome Metrics

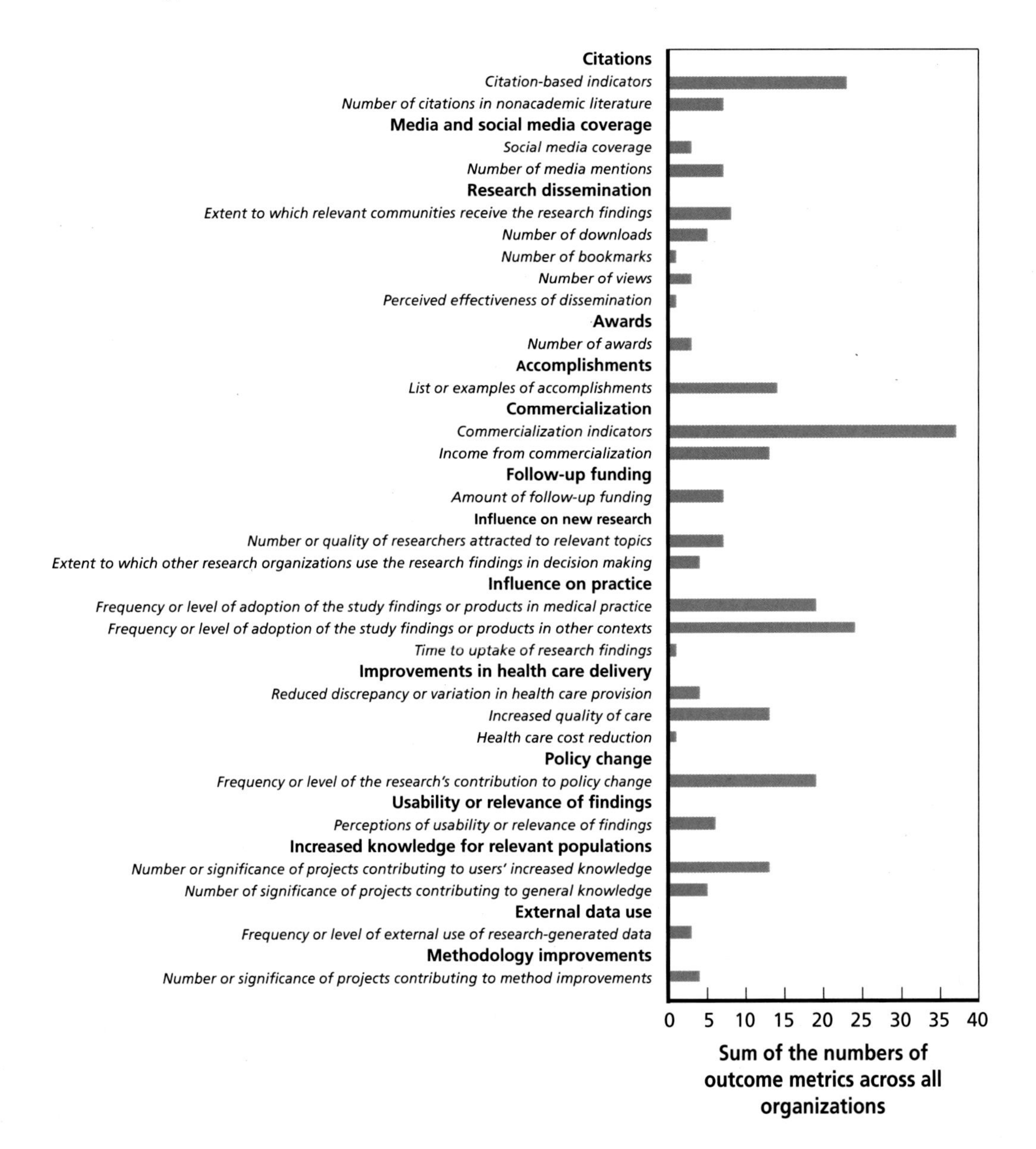

or articles; (2) variation in retrieved data based on data sets and sources used; and (3) gaps in information on social media users, giving rise to such concerns as the presence of automated bots or problematic categorization of social media (e.g., research versus the lay community). Ultimately, a review by Sugimoto et al., 2017 concluded,

> stakeholders that supply and demand altmetrics and the use of social media in academia must be cautious not to take the shadow for the substance, where measurable traces of research activities and impact become more important than the activities themselves. (p. 2051)

Other outcome metric types capture achievements beyond citations and references. *Number of awards* refers to prizes, honors, and recognition awarded to researchers, projects, and portfolios. *Narratives of accomplishments* refers to situations in which agencies select notable examples of outcomes, without any prescription for what form the achievements should take. As such, this type of metric can be understood as consisting of selected success stories or case studies of main accomplishments.

Other types of metrics are more prescriptive regarding the nature of the outcomes in question. It would be erroneous, however, to view the general accomplishment narratives and metric types that are more concrete as mutually exclusive. Rather, it is likely that the types of information that some agencies collect and on which they report as part of their general accomplishment narratives could be used to inform some of the metrics that are more prescriptive. For instance, it is likely that, as part of creating general outcome narratives, some agencies collect data on accomplishments in the form of influence on practice or policy, both of which are also listed as stand-alone metric types.

Among these, three key categories emerge: changes in practice, changes in health care delivery, and changes in policy. Changes in practice can be measured by the frequency or level of adoption of the results from the underlying research in medical practice, as well as by practitioners in other contexts. One metric type in this category examines the speed with which research findings are taken up. Research outcomes take the form of improvements in health care delivery. Metric types in this category include trends and reductions in discrepancy and variation in health care provision, as well as trends and increases in quality of care. The latter metric type could take a variety of forms; identified individual metrics under this metric type include patients' satisfaction with care; perceptions of quality; trends in the volume of inappropriate, unnecessary, and ineffective care; and trends in health care–acquired conditions. Reductions in health care costs are another metric type falling under the broader category of improvements in health care delivery. Finally, changes in policy represent an oft-desired outcome of research activities and can be measured by both the frequency and the level or significance of research contribution to observed policy change. In concrete terms, applicable individual metrics include numbers of clinical guidelines, regulations, standards, and other policy documents that refer to the underlying research. Narrative assessments of whether the research informed policy (e.g., informed by stakeholder testimonies) are also an option. Such policy changes can provide a pathway to change in practice, amplifying the impact of the motivating research.

In other categories, a series of metrics and metric types relate to commercialization. These include numbers of licenses (possibly broken down by, for example, whether they are new or whether they are exclusive), time taken to grant licenses, numbers of various technology transition indicators (such as cooperative research and development agreements, nondisclosure agreements, and material transfer agreements, as well as indicators about the partner agencies, such as the numbers of small businesses involved), numbers of spin-offs, numbers of start-ups, and the amount of capital invested in new companies. Another group of individual commercialization-related metrics examines income from commercialization activities (e.g., in the form of licensing or royalty income or sales and other financial indicators associated with developed products).

Another type of outcome relates to the ability to generate follow-up funding—for instance, in the form of the numbers of projects that attracted follow-up funding, the numbers of new initiatives or projects, or the amount of follow-up funding. Two outcome metric types capture influence on new research. One type examines the numbers and quality of researchers attracted to a particular topic (e.g., through the formation of new research initiatives or submission of research proposals); the other type looks at the extent to which other research organizations use the underlying research (as demonstrated by, for example, citations in patent applications submitted by other organizations).

One category of outcome metrics examines the usability and relevance of research findings as perceived by various stakeholder groups, and another one captures whether the underlying research has led to increased knowledge for relevant populations. With respect to the latter category, one metric type relates to increases in the knowledge of the intended users of research (e.g., medical practitioners), while another relates to the general population's knowledge.

The remaining two metric types cover the frequency or level of external use of research-generated data (e.g., measured by the numbers or funding amounts of projects that utilize data created as part of the underlying research) and numbers or significance of instances in which the underlying research resulted in improvements in methodology (as measured by, for example, the numbers of applications of newly generated methods).

Assessment

Not surprisingly, outcome metrics effectively demonstrate the trade-offs associated with the use of various downstream metrics (see Table 3.4). Some metrics, such as various commer-

Table 3.4
Assessment of Outcome Metrics

Metric	Reliance on Expert Judgment	Data Burden	Attribution Issues
Citations			
Citation-based indicators	Low	Low	Low
Number of citations in nonacademic literature	Low	Low	Low
Media and social media coverage			
Social media coverage	Low	Low	Low
Number of media mentions	Low	Low	Low
Research dissemination			
Extent to which relevant communities receive the research findings	Low	Medium	Low
Number of downloads	Low	Low	Low
Number of bookmarks	Low	Low	Low
Number of views	Low	Low	Low
Perceived effectiveness of dissemination	Low	Medium	Low
Awards			
Number of awards	Low	Low	Medium
Accomplishments			
List or examples of accomplishments	High	Medium	Low

Table 3.4—Continued

Metric	Reliance on Expert Judgment	Data Burden	Attribution Issues
Commercialization			
Commercialization indicators	Low	Low	Medium
Income from commercialization	Low	Low	Medium
Follow-up funding			
Amount of follow-up funding	Low	Low	Medium
Influence on new research			
Number or quality of researchers attracted to relevant topics	Medium	Medium	Medium
Extent to which other research organizations use the research findings in decision making	Medium	High	Medium
Influence on practice			
Frequency or level of adoption of the study findings or products in medical practice	Medium	Medium	Low
Frequency or level of adoption of the study findings or products in other contexts	Medium	Medium	Low
Time to uptake of research findings	Medium	Medium	Low
Improvements in health care delivery			
Reduced discrepancy or variation in health care provision	Medium	Medium	High
Increased quality of care	Medium	Medium	High
Health care cost reduction	Medium	Medium	High
Policy change			
Frequency or level of the research's contribution to policy change	Medium	Medium	Low
Usability or relevance of findings			
Perceptions of usability or relevance of findings	Low	Medium	Medium
Increased knowledge for relevant populations			
Number or significance of projects contributing to users' increased knowledge	Medium	Medium	Medium
Number or significance of projects contributing to general knowledge	Medium	Medium	Medium
External data use			
Frequency or level of external use of research-generated data	Low	High	Medium
Methodology improvements			
Number or significance of projects contributing to method improvements	Medium	Medium	Medium

cialization indicators, might require relatively limited judgment and might be relatively easy to collect. They might not provide a complete picture of the achievements of the underlying research because, in the absence of further corroborating information, they can be understood as proxy measures for outcomes and impacts generated by the commercial partner. By contrast, metrics related to influence on practice might be able to effectively capture the contribution that the underlying research has made to any observed outcomes. However, data necessary to inform such metrics might be more burdensome to collect (because, for example, of the need to consult external sources of information) and might require a higher degree of expert judgment (e.g., when assessing the unique contribution of the underlying research and its significance). Furthermore, some of the types of metrics identified here could give rise to methodological and interpretation issues that, in some instances, can limit the ability to accurately capture the results of the research or fully attribute claimed results to the underlying research. These considerations pertain to some citation-based indicators, as well as measures looking at engagement with the underlying research via alternative platforms.

Impact Metrics

Metric Types

Figure 3.9 presents an overview of identified types of impact metrics. Numerous agencies attempt to measure some form of long-term health–related outcome (e.g., changes in mortality, morbidity, functional status, incidence of specific conditions, life expectancy, quality of life); others examine trends in the utilization of health care or incidence of health risks.

The set of impact metric types also includes a series of measures of cost-effectiveness and economic returns. Applicable cost-effectiveness and unit-cost indicators include the costs per product (e.g., per publication, per presentation), as well as various budget components (e.g., overhead costs as a share of total funding, support staff–to-researcher ratio). Individual metrics of ROI include social, internal, or adjusted rates of return; benefit-to-cost ratios; and net present value.

Figure 3.9
Identified Types of Impact Metrics

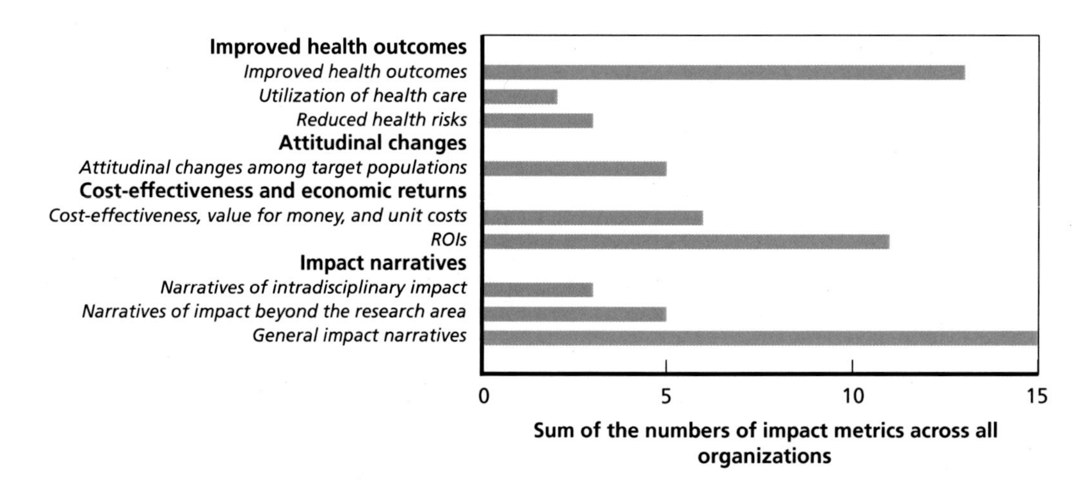

Last, a large number of agencies collect impact narratives, most typically in a general and unspecified form, although some also explore any interdisciplinary impacts and impacts extending beyond the research area in question. As with narratives utilized for the assessment of outcomes, one should not view impact narratives and the previous groups of impact metrics as mutually exclusive. Similarly, it is likely that at least some impact narratives will incorporate information that could, in another context, be used to inform impact metric types that are more specific.

Assessment

Table 3.5 illustrates the range of trade-offs associated with impact metrics. As discussed in Chapter Two, one of the biggest challenges in impact measurement is attribution issues. One way to address this challenge is to construct contribution stories that would form part of impact narratives and capture the link between the underlying research and the observed or claimed impacts. The collection of necessary evidence, particularly with respect to the significance of the underlying research for the observed impacts, can be burdensome. By contrast, data on health outcomes among populations of interest might be comparatively straightforward to collect, although, in the absence of further data, they will be subject to substantial attribution issues. An important caveat to the assessments provided below is that all three criteria are very context dependent and will depend on how the metrics in question are actually applied. To illustrate, as also discussed in Chapter Two, the burden associated with the development of narratives might depend on what data are already being collected and how much additional analysis is required. Furthermore, some complex nonnarrative assessments (e.g., ROI calculations or cost-effectiveness analyses) could be more burdensome than relatively simple narrative-based assessments (e.g., a single case study). Ultimately, we reiterate, the values

Table 3.5
Assessment of Impact Metrics

Metric	Reliance on Expert Judgment	Data Burden	Attribution Issues
Improved health outcomes			
Improved health outcomes	Medium	Medium	High
Utilization of health care	Medium	Medium	High
Reduced health risks	Medium	Medium	High
Attitudinal changes			
Attitudinal changes among target populations	Medium	High	High
Cost-effectiveness and economic returns			
Cost-effectiveness, value for money, and unit costs	Medium	Medium	Medium
ROIs	Medium	High	Medium
Impact narratives			
Narratives of intradisciplinary impact	High	High	Low
Narratives of impact beyond the research area	High	High	Low
General impact narratives	High	High	Low

presented in Table 3.5 are meant as a heuristic relative assessment and might not capture all possible combinations and forms of assessment.

CHAPTER FOUR

Conclusions for the Psychological Health Center of Excellence, the Defense and Veterans Brain Injury Center, and the Defense Health Agency

Summary of Main Points

There is no one-size-fits-all approach to portfolio assessment, and practices observed across research organizations vary accordingly.

Innovative work is taking place across research organizations in terms of building evaluation capacity and systems, as well as developing new tools, which suggests potential for cross-organizational learning.

Using its tracking system implemented in 2017, the former DCoE appeared to collect data on upstream metrics that are comparable to those that other research organizations collect.

By contrast, the former DCoE showed less focus on outcome and impact metrics.

Reflecting on these findings, we propose the following recommendations to DCoE's successors:

- Review the utility of all currently collected upstream data.
- Incorporate downstream metrics in assessment efforts, possibly via an incremental approach, ensuring a balanced mix of individual metrics to mitigate associated trade-offs.

This chapter presents conclusions drawn from the rapid review of both published material on research-portfolio evaluation and consultations with people responsible for research portfolios in a variety of agencies and other organizations. Although emphasis is placed on potential relevance to the remaining components of (the PH CoE [PHCoE] and DVBIC) and successors to DCoE, including elements of the DHA concerned with research administration (beginning with the Research and Development Directorate [J-9]), the conduct of the project demonstrated the broad interest in the nature and utility of research-portfolio metrics.

Big-Picture Observations

Our research for this project resulted in a database and taxonomy of research-portfolio metrics pegged to the stages of a research-portfolio logic model, as described in Figure 1.1 in Chapter One. Research-supporting agencies and organizations will have different emphases on those stages. The DHA, a relatively new agency established in 2013 to improve coordination and governance across the MHS (MHS, undated), takes particular interest in the downstream stages of impacts and outcomes of research, given its focus on the health, well-being, and readiness of people in the military. In principle, the applied and translational nature of the research supported by the former DCoE avoids the uncertainties about impacts and outcomes associated with basic research, but our research illuminated how the choice of metrics used can either reinforce or diminish the uncertainties that apply to more applied (as well as basic research) portfolios.

No metric is universally useful. The suitability of individual approaches to research evaluation depends on the circumstances and needs of a given organization because there are trade-offs associated with the selection of research assessment approaches. Metrics based on publications (discussed in "Output Metrics" in Chapter Three) might be common, but they are subject to shortcomings and criticisms that limit their usefulness and sometimes result in misuse. As concluded in Guthrie, Wamae, et al., 2013, multimethod approaches could be most likely to offer the desired level of robustness and meet the purposes of research evaluations.

Given the preponderance of available frameworks and tools for the evaluation of research portfolios, it is not surprising that the ways in which research organizations go about assessing their portfolios vary significantly (in addition to the material we collected, see, e.g., Bozeman and Rogers, 2001, and Ruegg, 2007). With respect to federally funded research, the Congressional Budget Office noted that metrics and evaluative prioritization vary by agency, agency mission, and research scope (Congressional Budget Office, 2007). In a broader context of agency performance, CEP observed that the scope and degree of institutionalization of evaluation and evidence-building activities varied across federal agencies and departments. Factors contributing to this situation that CEP identified include varying degrees of capacity across agencies and departments, as well as competing pressures for resources (CEP, 2017). Similarly, and specifically in the context of DoD health research, a 2017 Defense Health Board report notes the complexity of the management arrangements for the Defense Health Program research, development, test, and evaluation program (Defense Health Board, 2017).

Multiple authors have noted that there is no standard use of *program*, *project*, or *portfolio* across research organizations, which is itself a subject of research (Wallace and Rafols, 2015; Srivastava, Towery, and Zuckerman, 2007). This point was also stressed by consulted stakeholders, who invariably agreed that one defining characteristic of a portfolio is the fact that it brings together a body of research working toward a common goal. In addition, 13 interviewees highlighted that there can be multiple definitions of *research portfolio* applicable within a single agency, depending on the desired level of aggregation. Some agencies organize their research in individual programs, which might be structured around research topics or methods. In these situations, each program has its own research portfolio, and, in aggregate, those portfolios form the research portfolio of the organization as a whole. Agencies might further disaggregate individual programs into various lines of research (e.g., according to the stages of the clinical care continuum), which might also be considered research portfolios in their own right. Aspects that interviewees mentioned as ways to organize portfolios included research topics, types of science, types of investigator, and types of method, as well as types of funding and funding sources.

Everyone consulted for this project who commented on the external role of research metrics acknowledged that metrics can help to explain to those outside of the activity or entity how and why research is being done. Decision making—within an agency or organization or at the highest levels of government that decide on budgets—requires justification, to which metrics can contribute. This reality is captured by the four As introduced in Chapter One: accountability, analysis, allocation, and advocacy. It was also implicit in a DHA J-9 symposium organized in October 2017 that focused on ROI,[1] although interviewees occasionally cautioned against

[1] The purpose of the symposium was for subject-matter experts to exchange practices and experience with respect to assessing ROI in the domain of research and development. The event consisted of four panels representing the following stakeholder groups: DoD, industry, other federal government, and universities and nonprofits (MHS, 2017).

an emphasis on ROI because it can compound the difficulties that research evaluation already presents. Stakeholders interviewed for this project confirmed the reasons for evaluating portfolios discussed above and shared the following uses of such research assessments: (1) making the case for the organization's existence, activities, and funding to regulators and senior leaders; (2) detecting any procedural problems that might arise; (3) communicating a story about how and why research is being done; (4) analyzing the state of a long-running research area; and (5) uncovering the impact of research funding. (These, of course, correspond to the four As.)

All interviewees who commented on the topic appreciated that externally driven measurement imperatives had the benefit of ensuring that measurement happens. Although it was common for stakeholders to lament that some data requests seem arbitrary and excessive, consuming resources intended for the research itself, in some contexts, evaluations are commissioned for outside evaluation experts to perform after the research is complete. Examples of such evaluations include Shoemaker et al., 2007, and Abt Associates, 2012; more generally, collaborations with external partners to boost evidence-building capacities is one of the mechanisms identified in the final report by CEP (CEP, 2017). Where technology transfer is a plausible outcome, analyses of the economic impacts can be done. Retrospective appraisal can help to gauge whether the right data have been requested (and, ideally, aim future requests). Because it can be hard to connect research to its consequences, interviewees who could do so developed narratives explaining how research was done and what it accomplished—they connected the dots that can be measured with finer-grained metrics through a story accessible to lay, as well as expert, audiences.

Numerous research organizations have dedicated infrastructure for the evaluation of their research activities, including at the portfolio level and lower levels of aggregation. A notable example is NSF's Evaluation and Assessment Capability unit, established in 2013 explicitly for the purposes of program evaluation. The unit has been involved in providing support and resources in the areas of data collection, management, and integration, and it has played a role in "establishing mechanisms for NSF-wide leadership and coordination in program and portfolio evaluation" (NSF, undated [a]). Similarly, in 2011, NIH established its Office of Portfolio Analysis (OPA) in its Division of Program Coordination, Planning, and Strategic Initiatives, which itself is part of the Office of the Director. OPA's mission is to coordinate portfolio-analysis activities across all of NIH and help decision makers and research administrators evaluate and prioritize current, as well as emerging, research areas (NIH, 2018). Numerous federal agencies have also established dedicated technology transfer offices (see, e.g., FDA, 2018; NIST, 2018b; U.S. Department of Agriculture, undated; and NASA Technology Transfer Program, undated), which, as part of their duties, collect data on a variety of measures related to the uptake of their research products.

In addition to carrying out or supporting internal evaluations and assessments, resources and capabilities can also be deployed for developing new evaluation tools and methods. For instance, in FY 2015, NSF convened a Portfolio Analysis Taskforce, which has informed the creation and implementation of new tools. Results from that effort are techniques for the mining and visualization of textual data (NSF, 2016). Similarly, NIH's OPA has been working on the development of new data tools and methodologies. A recent example of such endeavors is the introduction of the RCR, a novel bibliometric indicator of scientific productivity developed by NIH researchers (Hutchins et al., 2016). Research assessments and evaluations can also benefit from improvements to the underlying data (e.g., through the development of common reporting systems and data repositories).

Finally, it is important to remember that measurement planning and analysis tend to happen within agency and organization silos, although the work of other entities can inform evaluation efforts of a given research-supporting organization. Some knowledge-sharing can come through the work of organizations that focus on both comparisons and evaluations. As discussed earlier, GAO has published a series of reports on research evaluation and performance assessment practices of federally funded organizations (GAO, 2015a; GAO, 2015b; and GAO, 2016), as have the National Academies (Institute of Medicine, NAS, and National Academy of Engineering, 2001; National Research Council, 2014). The National Academies have also organized various fora intended to further existing knowledge on the topic of research evaluations (e.g., National Academies of Sciences, Engineering, and Medicine, 2017b; National Academies of Sciences, Engineering, and Medicine, 2017a) and have conducted evaluations of other research entities (National Academies of Sciences, Engineering, and Medicine, 2016). Our interviewees' interest in seeing our results reflects how limited the visibility of metric development and use across agencies and organizations is outside of contexts in which mechanisms exist for interagency coordination (such as the National Coordination Offices for the Networking and Information Technology Research and Development program and the National Nanotechnology Initiative, both under the aegis of the National Science and Technology Council [Office of Science and Technology Policy, undated]). The DHA could provide that kind of mechanism within the MHS.

Observations for Successors of the Defense Centers of Excellence for Psychological Health and Traumatic Brain Injury

In its last year, DCoE developed a data dictionary and piloted a research-tracking system within its components. The tracking system collected data in six categories corresponding to different performance-management objectives (see Table 4.1). The associated data dictionary

Table 4.1
Summary of Defense Centers of Excellence for Psychological Health and Traumatic Brain Injury Tracking System Components

Variable Category	Management Objective	Example of Variable
Administrative data	Track research studies from an administrative perspective.	Grant number, title, PI
Research-portfolio description	Describe DCoE's research portfolio.	Type of study, topic area, population of interest
Research finances	Describe the finances that support DCoE research.	Type of funding, volume of funding
Research agreements	Describe DCoE collaborations and completed agreements that support DCoE research.	Type and number of agreements associated with research studies, patents
Research risks	Identify risks to ongoing DCoE research.	Risk probability, risk impact
Literature contributions	Identify DCoE research's contributions to the state of the science.	Publications, presentations, reports

SOURCE: Data dictionary that RPM provided to the authors.

(reproduced in full in Appendix F) specifies what variables are included in the data-tracking system, defines those variables and their possible values, and determines the time and frequency of relevant data collection.

Although the dissolution of DCoE renders some aspects of that system obsolete, consultations undertaken during this study suggest that the system continues to provide an analytically useful reference point for surviving components or successors, including the DHA J-9. The following findings are perhaps best generalizable to PHCoE and DVBIC but could have broader applicability and implications for the DHA as well.

The Psychological Health Center of Excellence's and Defense and Veterans Brain Injury Center's Upstream Metrics Appear in Line with Common Practice

Data collected by (what was then) DCoE's tracking system that can be used to inform upstream portfolio assessment metrics (i.e., inputs, processes, and outputs) appear broadly in line with those reported by other research-supporting agencies and organizations. Although there are examples of metrics that other organizations use that cannot be informed by DCoE's tracking data, these do not represent critical omissions. There do not appear to be notable gaps in the data collected on downstream metrics. Although it is possible to argue that some data collected by DCoE are not strictly necessary to construct meaningful performance metrics, there might be merit in retaining all of the components if they prove necessary to comply with requests for information, whether for GPRA and GPRAMA compliance or for other research-administration purposes. Decisions about retention of tracked data categories should factor in the costs or burden associated with data collection relative to the benefits suggested by likely use of the data. In this regard, we reiterate that interviewees cautioned against the trend toward asking for data indiscriminately as imposing both direct and opportunity costs.

The Psychological Health Center of Excellence and the Defense and Veterans Brain Injury Center Seem to Lag in Outcome and Impact Measures

In contrast with its efforts to collect data on upstream metrics, DCoE's data tracking does not capture information beyond immediate research outputs, such as publications and patents. This is a major point of departure from the practice in research agencies and organizations covered in our rapid review. DCoE's tracking does not appear to capture systematically what happens with the results of its research once completed. As a consequence, its tracked data would yield only limited ability to comment on the extent to which research has impacts on or contributed to improvements in the key domains of service members' readiness and health.

The limited collection of data on downstream metrics in the pilot tracking system might reflect any combination of three considerations. First and foremost, collection of data on the uptake of research and, by extension, its outcomes and impacts, could have been considered as extending beyond DCoE's mission. If the mission is conceived primarily as identifying existing knowledge gaps and contributing to their closure, the collection of data on and assessments of knowledge production might be sufficient. Second, the fact that DCoE research activities resulted primarily in knowledge products might make collecting data on their uptake and resulting outcomes and impacts more challenging. Two interviewees stressed the difficulty of evaluating knowledge products, sometimes drawing comparisons with research leading to new materiel, the uptake of which is considered easier to monitor. And third, as noted earlier, the collection of data, particularly on downstream measures, is potentially very resource-intensive, and DCoE's components or successors might find that data collection is not the best use of the resources available.

Nevertheless, the position of DCoE's components (and, by extension, other entities under the aegis of the DHA J-9) within the MHS offers opportunities for outcome and impact measurement that might not be available to non-DoD organizations. First, the research of such entities has a clearly defined population of beneficiaries that is comparatively well documented, which might allow for a more efficient collection of data. Similarly, that research is intended to inform and benefit a closed health system with a well defined set of providers and facilities. These factors might help alleviate some of the data-collection burdens that other research organizations face.

Ending Ongoing Research Efforts Early Might Not Be Easy or Desirable

Questions emerged during consultations with DCoE officials throughout the research project about whether portfolio performance metrics could be useful for potential decisions about discontinuing ongoing research efforts—for example, if funding or institutional priorities shifted. Key informants consulted in the course of the study invariably stressed that there are strong obstacles to an early termination of research efforts that are already under way. These stem from legal, as well as cultural, factors and, as interviewees generally agreed, might be more pronounced in public research organizations than in their private counterparts.

Projects can be and are terminated on the grounds of nonperformance (e.g., in situations in which the research team fails to meet stipulated milestones established at the beginning of the research project). Like those of other research-supporting agencies and organizations, components of DCoE collect the necessary data on research progress and milestone attainment. Three interviewees explicitly noted that early termination typically occurs only in instances of very serious nonperformance or noncompliance with specifications. Some interviewees acknowledged that sometimes expectations about the effectiveness of ongoing research projects can change; those interviewees emphasized that predicting whether a study was on track to achieve impact is inherently difficult. This difficulty militates against discontinuing a study that was formally meeting all agreed milestones. In fact, three interviewees specifically warned against attempting to judge a project's achievements too early in the research. A possible exception would be a situation in which there has been a material change in the context in which the research project is taking place (e.g., a departure of a key collaborator, adopter, or stakeholder who was instrumental in the plans for the uptake of the research results).

As all interviewees who commented on the topic affirmed, considerations surrounding the likely impact of individual projects are part of the decision-making process on what project proposals will be approved for funding. For that process, the expected outcomes and impacts are formulated; depending on the context, an assessment of the likelihood of their achievement, as well as the factors affecting this likelihood, can be made. Consequently, it might be possible to periodically reassess the estimated likelihood of achieving impact throughout the course of the research project in light of new developments and progress data, which might, in turn, help inform decisions as to which projects might be candidates for early termination.

Having some degree of flexibility built into research contracts to adapt to changing circumstances has been found in existing literature to be effective in delivering positive outcomes and impacts (Wooding et al., 2004). This suggests that a more flexible framework and less rigid milestones, along with a communicative relationship with the funder, might be a better way to introduce this adaptiveness, as opposed to trying to end projects that do not conform to the original plans or that need to change in response to changing priorities.

The units of analysis in this discussion of early discontinuation are primarily individual research projects. Where stakeholders commented on the possibility to recalibrate the composition of research portfolios as a whole, they pointed to the proposal-review stage and decisions about nonrenewal or nonextension of existing research projects as the two most important avenues. In the context of recalibrating portfolios, we reiterate the existence of literature on decision making and portfolio analysis, which could offer relevant lessons and suggestions (see, e.g., Linton and Vonortas, 2015; Wallace and Rafols, 2015; Klavans and Boyack, 2017; and Rafols and Yegros, 2018).

Current Needs and Possible Actions

Building on the findings presented earlier, we identified a series of possible actions for the consideration of PHCoE, DVBIC, and the DHA more broadly. These are the product of the research team's analysis of the inputs we collected and the applications of our expert judgment. Because the original project sponsor no longer exists in its original form, recommendations are intended to make sense in the new context, which is still evolving.

Possible Action with Respect to Upstream Metrics

Review the value of currently collected data on upstream metrics (inputs, processes, and outputs). As discussed earlier in this chapter, the current extent of data collection by the former DCoE appears broadly in line with the practice reported by other research organizations. Because the collection of data can be costly and burdensome, it might be useful to review whether the benefits of all currently collected data outweigh the costs incurred in the process of their collection (and maintenance). In assessing the benefits of having a particular type of information available, it is possible that, for at least some metrics, their utility will lie primarily in responding to requests for information from external stakeholders (as opposed to informing performance assessments). This recommendation arises from the observation that the current scope of upstream data collection is extensive, whereas less is done with downstream metrics (especially outcomes and impacts), where resources could be used more effectively.

Identify opportunities for streamlining reporting requirements and activities. Some data burden associated with portfolio assessment is incurred not at the stage of data collection but rather during efforts to process and analyze the data. This challenge is exacerbated in situations in which different audiences of performance-reporting products ask slightly different questions or expect data to be presented in slightly different forms. In the absence of a robust and easily searchable central repository of information, this variation in reporting requirements and requests can result in duplicative work, as well as direct and opportunity costs. Existing reporting arrangements and needs should be reviewed to ascertain whether opportunities might exist for greater efficiencies. As discussed in Chapter Two, some agencies and organizations make great use of central information systems to facilitate the production of analytical outputs and responses to ad hoc information requests. This practice could offer valuable transferable lessons for the DHA. This recommendation arises from concerns expressed by interviewees in DoD entities about reporting burdens and the positive examples gleaned from non-DoD entities of the use of central information systems.

Possible Action with Respect to Downstream Metrics

Incorporate outcome and impact measurements in tracking and assessment processes. During this rapid review, we found a contrast in practice between components of the former DCoE and other DoD research-supporting agencies and organizations on one hand and those operating in civilian contexts on the other. Specifically, although practices across organizations vary, the former DCoE and other DoD entities addressed appear to focus on the measurement of outcomes and impacts less than their non-DoD counterparts do. To close this gap, it would be beneficial to investigate ways to introduce a more robust assessment of outcomes and impacts. To guide such an effort, it might be helpful to develop an impact measurement framework that would provide strategic guidance for the collection and analysis of outcome and impact data, as some non-DoD organizations (e.g., AHRQ and NIOSH) have done. This can be, in turn, informed by building a logic model capturing the presumed progression from research input and activities toward outcomes and ultimate impacts. Implementing such logic-model development and application is illustrated in Williams et al., 2009, and Landree, Miyake, and Greenfield, 2015, both of which involved the construction of logic models for evaluation efforts undertaken by NIOSH. Further guidance can be found in handbooks by Paul, Yeats, Clarke, Matthews, and Skrabala, 2015, and Savitz, Matthews, and Weilant, 2017. This recommendation arises from two observations: First, non-DoD entities have been doing more to measure research-portfolio outcomes and impacts; and second, among the DoD entities examined, there is an opportunity for measurement to be more systematic for research-portfolio outcomes and impacts, which appears from communication with the sponsor to be a goal within DoD.

Consider developing outcome and impact tracking and measurement in an incremental fashion. The implementation of outcome and impact measurement mechanisms could benefit from an incremental approach. It might not be feasible or practical to introduce a wide range of metrics at the same time; a more graduated approach might be most practical. To illustrate, in lieu of focusing on the ultimate impact of research on readiness and health (e.g., in the form of quantitative data on relevant disease burdens [and their trends] coupled with a narrative of the research's contribution to the observed trends), it might be more realistic to initially concentrate on selected intermediate outcomes. There are several advantages of such an approach. Measuring intermediate outcomes is likely to be less difficult and burdensome than measuring impacts. At the same time, the attainment of intermediate outcomes can serve as a reasonable proxy for the likelihood of achieving ultimate impacts. The collection of outcome data can also lay the foundation for any claims of the research's contribution to eventual longer-term impacts, should a decision be made to investigate these in the future. This recommendation reflects our judgment that wholesale change might not be practical and that experimentation with alternatives (informed by the examples collected and discussed in this report) might be beneficial.

Where even data on intermediate outcomes are currently not feasible to collect, it might be possible to start with collecting data on the dissemination and audiences for the research. In this context, it would be helpful to get a better understanding of the reach of the research (i.e., the number and breadth of stakeholders reached by the work), as well as, to the extent possible, the significance of the research (e.g., stakeholders' assessment of its usefulness). As with intermediate outcome data, dissemination-related data would be valuable in their own stead but could also provide the basis for any subsequent outcome and impact measurements further down the line.

Construct a balanced mix of metrics and determine how underlying data will be collected. As stressed at the beginning of this chapter, the purposes of research-portfolio assessments are likely best served by an approach that incorporates a mix of tools and methods. Therefore, in selecting what downstream metrics to use, it will be important to construct a basket of measures that will be able to mitigate any negative impacts stemming from the tradeoffs associated with the use of individual metrics. This selection should consider the purposes and audiences for these metrics to ensure that they balance the effort involved in their implementation with their utility to key stakeholders and for their intended use. This recommendation reflects lessons learned from the literature and the judgment of the research team that, in a world of research constraints and performance-measurement demands, there is an opportunity to make explicit choices about the metrics used at each stage represented by the logic model. These choices can help an organization achieve a better balance than what was observed and reported here without adding to reporting burdens.

The overview of metrics presented in Chapter Three provides a picture of existing practice, as well as an indication of the relative use of individual metric types. For instance, with respect to outcomes, a large number of organizations work to measure at least one of the three principal forms of change: (1) in practice, (2) in policy, and (3) in attitudes. It is likely that similar types of metrics would be relevant for any outcome measurement efforts by PHCoE, DVBIC, and the DHA more broadly. Furthermore, the three criteria used in our assessment of metrics—expert judgment, data burden, and attribution issues—represent a conceptual frame of reference that can be used to inform and refine the selection of new performance metrics, with the aim of finding an acceptable balance across all three dimensions. To assist any prioritization efforts on the part of PHCoE, DVBIC, or the DHA more broadly, Appendix B includes a suggested priority ranking of metrics at each stage of measurement.

Last, as part of this process of selecting downstream metrics, it will be necessary to consider the appropriate data-collection and assessment infrastructure. This includes such dimensions as (1) who will be responsible for the data collection, (2) the source of the data, (3) when (and how frequently) the data should be collected, and (4) what or who is required to process and analyze the data.

APPENDIX A

Variety in Evaluation Frameworks and Tools

To explore the range of approaches to research evaluation, Guthrie, Wamae, and colleagues conducted a comparative analysis of 21 international research-evaluation frameworks and their characteristics (Guthrie, Wamae, et al., 2013).[1] The authors identified a series of correlations between the identified characteristics of analyzed frameworks, which enabled them to make several key observations.

First, quantitative approaches can be used to produce longitudinal data without the need for interpretation or expert judgment. In addition, these approaches are relatively transparent. On the other hand, they can be relatively resource-intensive, at least at the initial stage of developing and implementing the framework.

Second, approaches that focus on learning and improvement (i.e., formative approaches) are generally suitable for evaluations spanning a range of areas. At the same time, they are not amenable to cross-institutional comparisons.

Third, some approaches require a strong degree of coordination and centralized input. Such approaches are not generally very suitable for use on a frequent basis.

Fourth, approaches with a notable degree of central input (in the form of either ownership or oversight) were found to tend to be more advanced in their implementation stage.

And fifth, some approaches are characterized by high levels of participant burden. Such approaches require that levels of participant expertise be sufficiently high (or it should be possible to provide training and build capacity as necessary).

Figure A.1 shows the potential trade-offs among approaches to research evaluation and how these relate to the underlying objectives of evaluation efforts, expressed by the four As: accountability, allocation, analysis, and advocacy. For instance, if the objective of research assessment is allocation of resources, some degree of comparison is likely necessary. By extension, this means that frameworks that could be used for such a purpose will likely be more summative than formative and not very comprehensive.

In addition to the variety of research-evaluation frameworks, a variety of evaluation tools could be employed to support the frameworks. Those include bibliometric analyses, economic analyses, data-mining approaches, surveys, logic modeling, document review, peer review, key-informant and stakeholder interviews, site visits, and case studies.

As they did with research-evaluation frameworks, Guthrie and her colleagues conducted a comparative analysis of the characteristics of individual research-evaluation tools (Guthrie,

[1] The work by Guthrie et al. has received international recognition (e.g., in the National Academies' study on furthering U.S. research efforts [National Research Council, 2014]. However, there are other reviews of existing research-evaluation frameworks [e.g., Penfield et al., 2014]).

Figure A.1
Typology of Research-Evaluation Frameworks

Transparent, longitudinal, quantitative, or free from judgment
Low up-front burden or qualitative
Formative, flexible, or comprehensive
Produces a comparison
Analysis
Accountability
Advocacy
Allocation

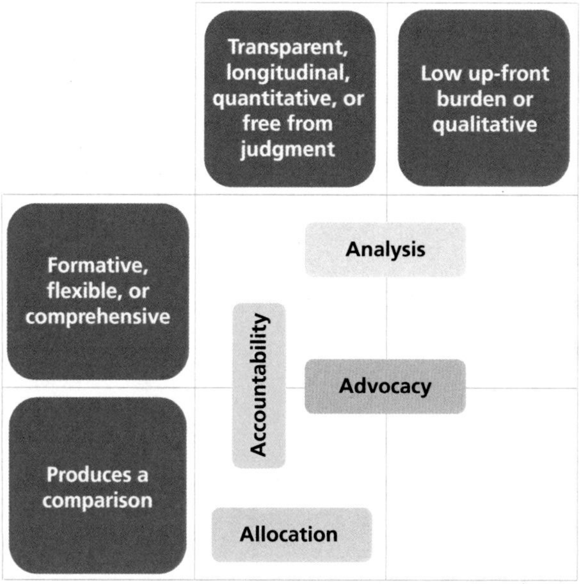

SOURCE: Guthrie, Wamae, et al., 2013, p. 9.

Wamae, et al., 2013). The authors identified two broad groups of tools. The first group consists of tools that are flexible and formative and that lend themselves well to cross-disciplinary analyses and comparisons. Examples of such tools include case studies and site visits, as well as interviews and document and peer review. The other group of tools are primarily quantitative in nature and free from expert judgment. They are more suitable for repeated and longitudinal use and are marked by a high degree of transparency. Examples of tools in this category include bibliometric and economic analyses, data mining, and surveys. Guthrie and her colleagues concluded that multimethod approaches could be most likely to offer the desired level of robustness and meet the purposes of research evaluations.

Research Evaluation Can Take Place at Various Stages

Table A.1 illustrates the difference in the covered stages of measurement in selected international research-evaluation frameworks (Guthrie, Wamae, et al., 2013). All frameworks capture output metrics, and most also include input and outcome metrics. By contrast, impact metrics, which are more challenging to incorporate, are less common. In addition to those four categories, the measurement of processes is also possible, although it was not explicitly covered as a stand-alone category in the table in Guthrie, Wamae, et al., 2013.

Research Evaluation Can Take Place at Various Levels of Aggregation

Guthrie and her colleagues examined variation in both the level targeted by research evaluation and how the unit of analysis is typically different with respect to collection and data

Table A.1
Stages of Measurement Captured by Various International Research-Evaluation Frameworks

Metric Type	CAHS (Canada)	ERA (Australia)	National Institute for Health Research (UK)	Productive Interactions (Several European Countries)	REF (UK)	STAR METRICS (U.S.)
Input	x	x	x		x	x
Output	x	x	x	x	x	x
Outcome	x		x	x	x	
Impact	x				x	

SOURCE: Guthrie, Wamae, et al., 2013.
NOTE: ERA = Excellence in Research for Australia.

reporting (Guthrie, Wamae, et al., 2013). Data are collected at a more granular level and subsequently reported either at the same level or aggregated to apply to a higher level of analysis. This is captured in Table A.2, which presents the various levels of aggregation utilized in six international research-evaluation frameworks selected for in-depth analysis for Guthrie, Wamae, et al., 2013.

Table A.2
Units of Analysis Used by International Research-Evaluation Frameworks

Unit	CAHS	ERA	National Institute for Health Research	Productive Interactions	REF	STAR METRICS
Research system	Reporting	Reporting	Reporting			Reporting
Field	Reporting	Reporting	Reporting			Reporting
Institution	Reporting	Reporting			Reporting	Collection
Department or program	Collection		Collection	Collection	Reporting	
Research group	Collection		Collection	Collection		
Project					Collection	
Researcher		Collection			Collection	

SOURCE: Guthrie, Wamae, et al., 2013.
NOTE: The cells indicate whether the organization uses that unit of analysis for data collection or data reporting.

APPENDIX B

Suggested Prioritization of Metrics

Tables B.1 through B.5 provide an overview of how then-DCoE's data-tracking system compares with the metric types identified as part of this rapid review and discussed in detail in Chapter Three of this report. Each table used for these comparisons presents an individual metric type and two assessment columns. The first assessment column denotes whether data

Table B.1
Suggested Priority Ranking: Input Metrics

Metric	Already Measured?	Priority
Grant application or decision		
Number or success rate of grant applications	No	2
Number of grant panels	No	2
Staff		
Number of researchers	Yes	1
Distribution of research specialties among PIs	No	1
Number of collaborations	No	2
Indicators of investigator quality or number of external affiliations	No	2
Other or miscellaneous input metrics (e.g., organizational prestige)	No	3
Funding		
Total amount of research funding provided by the organization	Yes	1
Internal funding by the portfolio or program	Yes	1
Total amount spent on the research topic	Yes	1
Comparison of budget allocation with that of other funders	No	2
Total amount of external funding	No	1
Amount of funding, by source	Yes	1
Facilities or tools used in research		
List of research infrastructure elements used in research activities	No	3
Frequency or volume of use of research infrastructure	No	3

Table B.2
Suggested Priority Ranking: Process Metrics

Metric	Already Measured?	Priority
Number of studies		
Number of projects funded	Yes	1
Number of collaboratively executed projects	Yes	2
Number of projects, by study type	Yes	1
Number of projects related to other research studies	No	2
Research topics		
Importance of the topic covered by the funded studies	No	2
Number of research projects, by research topic	Yes	1
Populations that the research covers, whether as the study focus or as participants	Yes	2
Alignment with the organization's goals	Yes	1
Milestone completion and progress tracking		
Milestone completion and progress tracking	Yes	1
Results reported back to study participants		
Extent to which results are reported back to study participants	No	2
Administrative burden		
Measures of administrative efficiency or burden	No	1
Other activities		
Other activities	No	2

that DCoE's tracker collected could be used for the metric type in question.[1] This analysis is based primarily on DCoE's data dictionary, shared with the research team. The second assessment column represents an indication of a priority ranking for each metric type. The ranking, ranging from 1 (highest priority) to 3 (lowest priority), serves to highlight metric types that might be most important to collect in a resource-constrained environment. This ranking was developed at an internal synthesis workshop we held that aimed to reflect on the evidence collected as part of this study and formulate conclusions and recommendations. The ranking displayed in this appendix is a product of the research team's deliberation and consideration of collected data, as well as of the team members' expertise in the fields of research and performance evaluation. Other analysts or stakeholders could make a case for a different ranking.

The objective of this effort was to apply the findings from the rapid review and our mode of thought to an existing mechanism—DCoE's data dictionary—to provide concrete illustration and, if desired, a basis for rapid implementation. The prioritization exercise suggests answers to the two following questions. First, with respect to inputs, processes, and outputs for which DCoE's data dictionary is broadly in line with practice elsewhere, are there any notable

[1] In a few instances, although the data tracking does not explicitly include a measure presented in the table, data collected for another measure could also be used to inform the original measure. These cases are denoted as *possibly*.

Table B.3
Suggested Priority Ranking: Output Metrics

Metric	Already Measured?	Priority
Total production		
Total number of outputs	Yes	1
Publications		
Number of publications (possibly organized according to program)	Yes	1
Number of peer-reviewed publications	Yes	1
Publication quality	Possibly	1
Number and quality of publication collaborations	Possibly	2
Presentations		
Number of presentations	Yes	2
Other research products		
Number of new devices or treatments	No	2
Number of new knowledge tools	Yes	1
Number of educational materials developed	No	2
Number of data sets	No	2
Number of websites and web-based products	No	3
Number of other research products	No	3
Training		
Number of training activities or mechanisms	No	3
Number of training beneficiaries or their feedback	No	3
Dissemination or outreach		
Number of dissemination or outreach activities or products	No	1
Public access to reports	No	2
Number of media-related products (e.g., press releases)	No	3
Patents		
Number of patents	Yes	1

gaps in the scope of data collection? And second, with respect to outcome and impact metrics (for which the data dictionary includes very few items), what would be the metrics most essential to add, particularly if one were to adopt an incremental approach to building an outcome and impact measurement framework?

With respect to the first question, we concluded that two metrics not included already would be useful additions to the existing upstream metrics: (1) a measure of administrative efficiency and (2) a measure of dissemination activities. The rationale for these two is that the DCoE data dictionary does not offer a measure that would enable (1) an assessment of efficiency, cost-effectiveness, or value for money or (2) an account of activities aiming to get

Table B.4
Suggested Priority Ranking: Outcome Metrics

Metric	Already Measured?	Priority
Citations		
Citation-based indicators	Possibly	1
Number of citations in nonacademic literature	No	3
Media and social media coverage		
Social media coverage	No	3
Number of media mentions	No	3
Research dissemination		
Extent to which relevant communities receive the research findings	No	1
Number of downloads	No	3
Number of bookmarks	No	3
Number of views	No	3
Perceived effectiveness of dissemination	No	2
Awards		
Number of awards	No	3
Accomplishments		
List or examples of accomplishments	No	1
Commercialization		
Commercialization indicators	Yes	1
Income from commercialization	No	1
Follow-up funding		
Amount of follow-up funding	No	3
Influence on new research		
Number or quality of researchers attracted to relevant topics	No	3
Extent to which other research organizations use the research findings in decision making	No	3
Influence on practice		
Frequency or level of adoption of the study findings or products in medical practice	No	1
Frequency or level of adoption of the study findings or products in other contexts	No	1
Time to uptake of research findings	No	2
Improvements in health care delivery		
Reduced discrepancy or variation in health care provision	No	1
Increased quality of care	No	1

Table B.4—Continued

Metric	Already Measured?	Priority
Health care cost reduction	No	1
Policy change		
Frequency or level of the research's contribution to policy change	No	1
Usability or relevance of findings		
Perceptions of usability or relevance of findings	No	2
Increased knowledge for relevant populations		
Number or significance of projects contributing to users' increased knowledge	No	2
Number or significance of projects contributing to general knowledge	No	2
External data use		
Frequency or level of external use of research-generated data	No	2
Methodology improvements		
Number or significance of projects contributing to method improvements	No	2

research results to intended audiences. The team ascribed lower priority to other measures not currently covered by the data dictionary tool, generally because the data dictionary already includes a related measure in the same category of metric.

Table B.5
Suggested Priority Ranking: Impact Metrics

Metric	Already Measured?	Priority
Improved health outcomes		
Improved health outcomes	No	1
Utilization of health care	No	2
Reduced health risks	No	2
Attitudinal changes		
Attitudinal changes among target populations	No	2
Cost-effectiveness and economic returns		
Cost-effectiveness, value for money, and unit costs	No	1
ROIs	No	1
Impact narratives		
Narratives of interdisciplinary impact	No	2
Narratives of impact beyond the research area	No	2
General impact narratives	No	1

With respect to downstream metrics, the suggested prioritization reflects the finding from the rapid review that the research-evaluation measures that are most fundamental revolve around the following questions:

- Who has used the research?
- What has the research achieved?
- Has the research led to the development of new products or ventures?
- Has the research resulted in changes in practice?
- Has the research resulted in changes in policy?
- Has the research contributed to improved outcomes of interest?

Metrics deemed to be the highest priority are those related to the uptake of the research (e.g., citations); commercialization indicators; changes in policy, practice, and (health) outcomes; and lists or narratives of accomplishments and results. Of course, this leaves out other measures, such as changes in attitudes or knowledge, which are assessed as lower priority but could be very important in their own right. Ultimately, because it is not practical or feasible to incorporate every measure, the prioritization in this appendix represents a suggestion for what choices to make.

APPENDIX C

Additional Data on Metrics

This appendix presents additional extractions from the database of metrics compiled for this study. These extractions capture the following types of disaggregation: (1) frequency of individual metrics, by broad metric category (analytical level 2); (2) frequency of metric types, by agencies' research focus; and (3) frequency of metric types, by type of organization.

Frequency of Metrics, by Broad Metric Category

Figure C.1
Identified Input Metric Categories

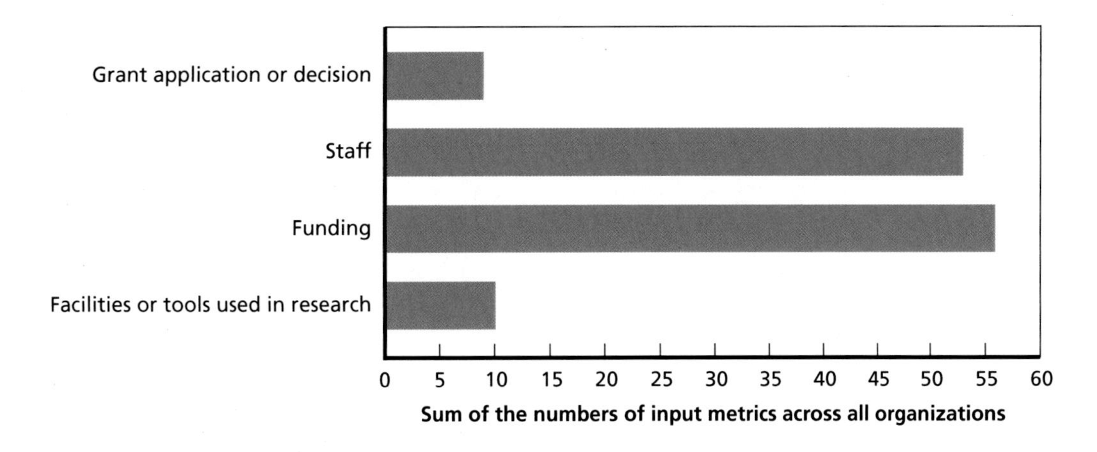

Figure C.2
Identified Process Metric Categories

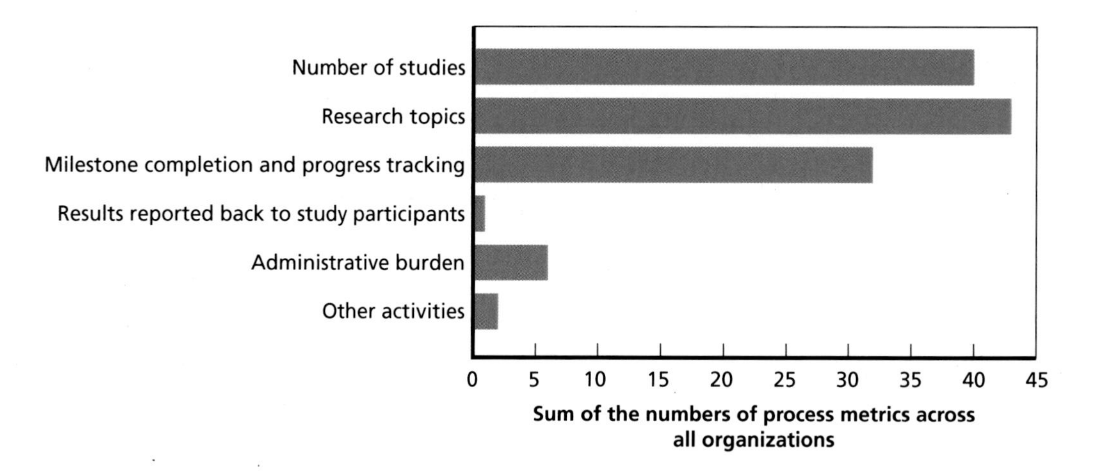

Figure C.3
Identified Output Metric Categories

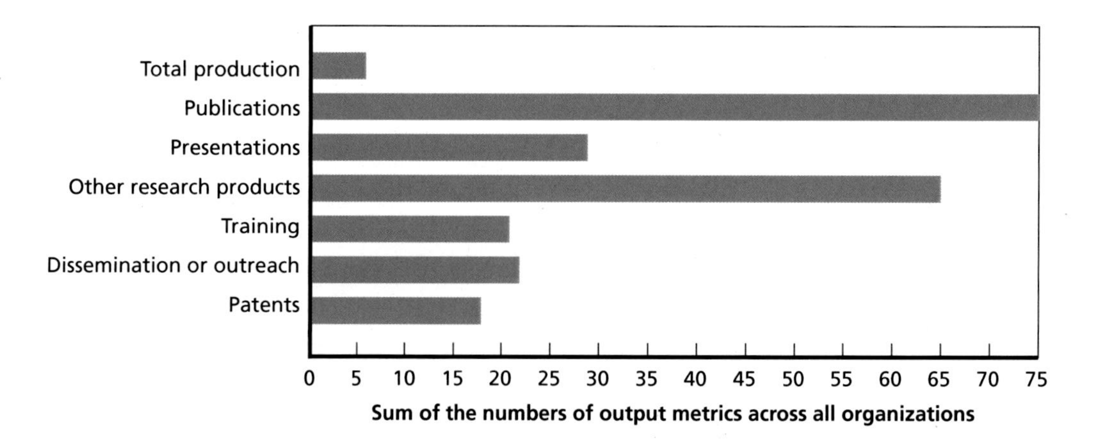

Figure C.4
Identified Outcome Metric Categories

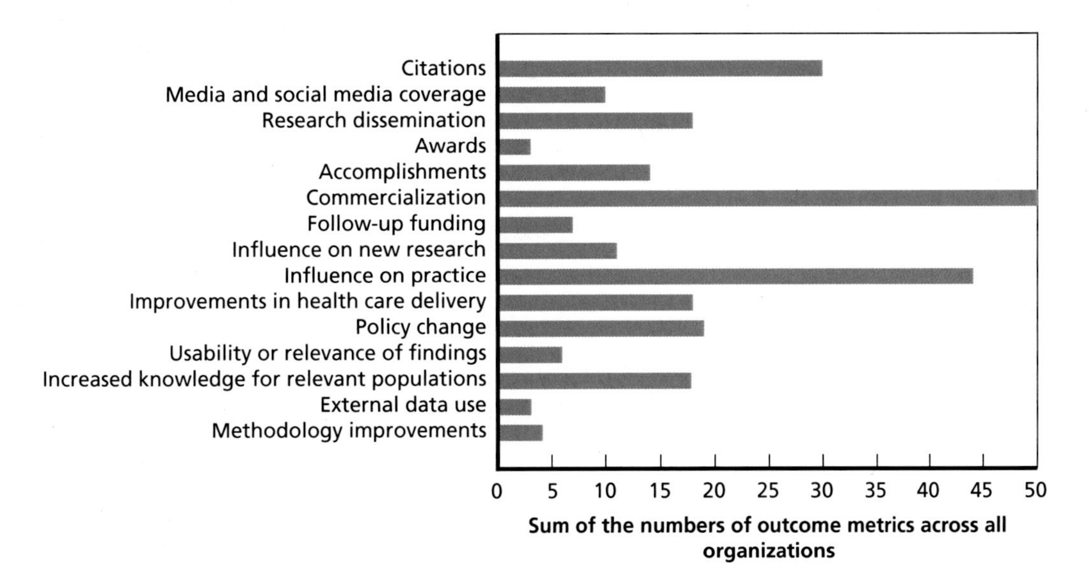

Figure C.5
Identified Impact Metric Categories

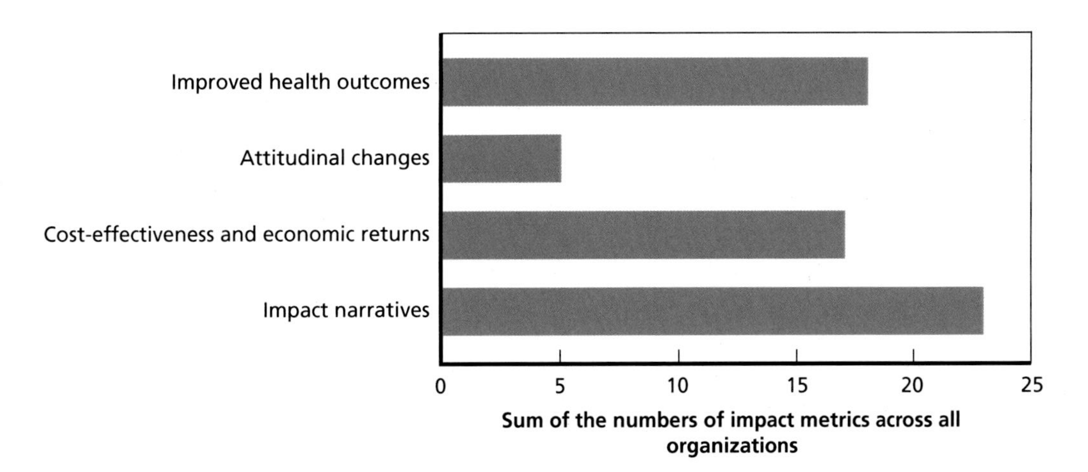

Frequency of Metric Types, by Agencies' Research Focus

Figure C.6
Input Metrics, by Type of Research

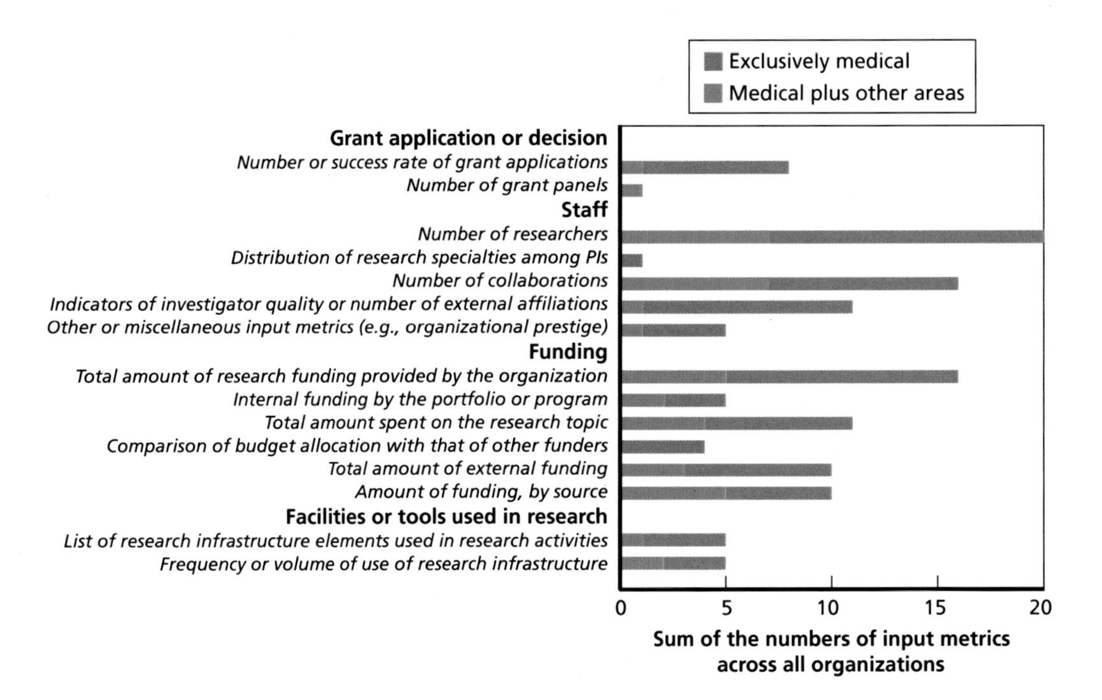

Figure C.7
Process Metrics, by Type of Research

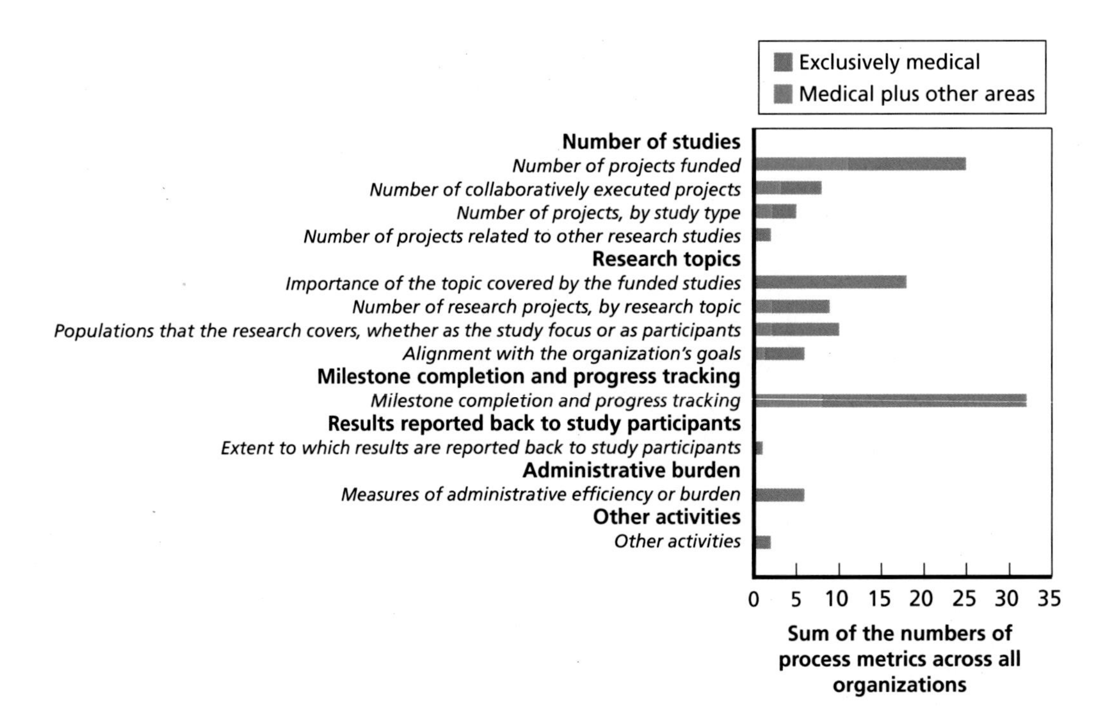

Figure C.8
Output Metrics, by Type of Research

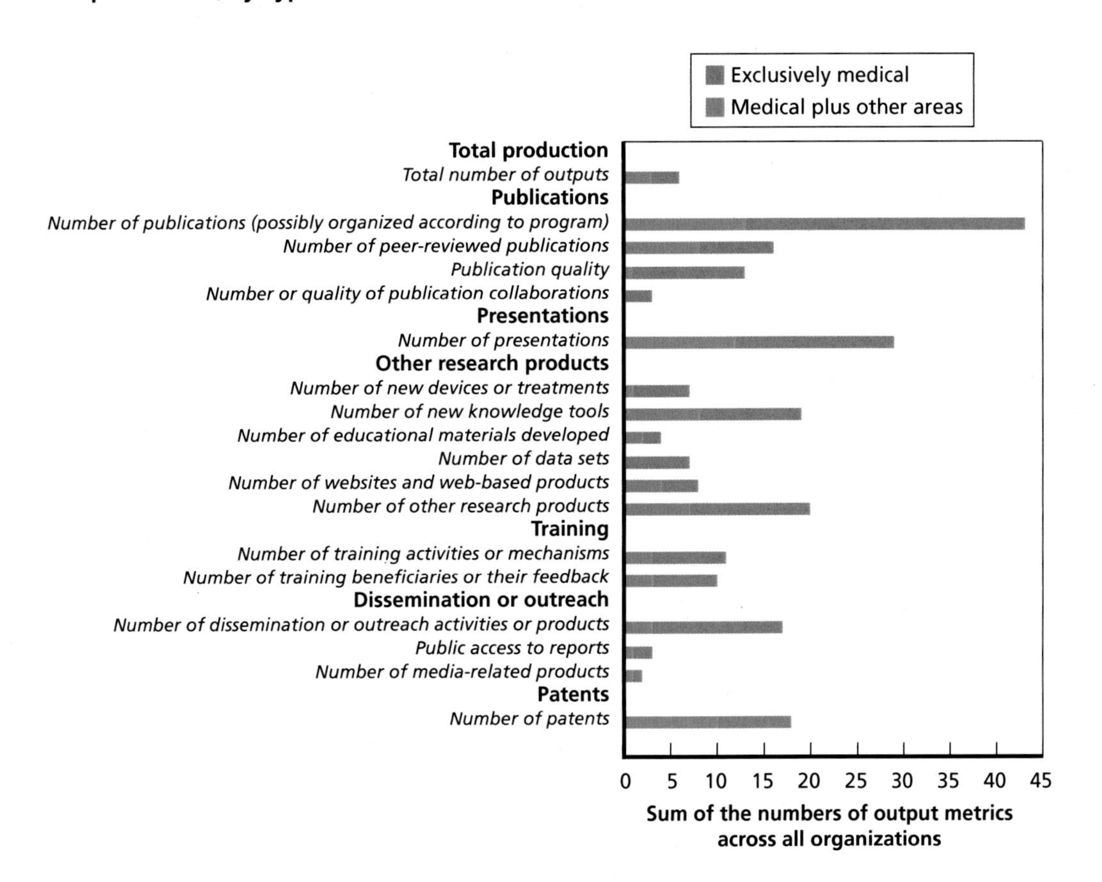

Figure C.9
Outcome Metrics, by Type of Research

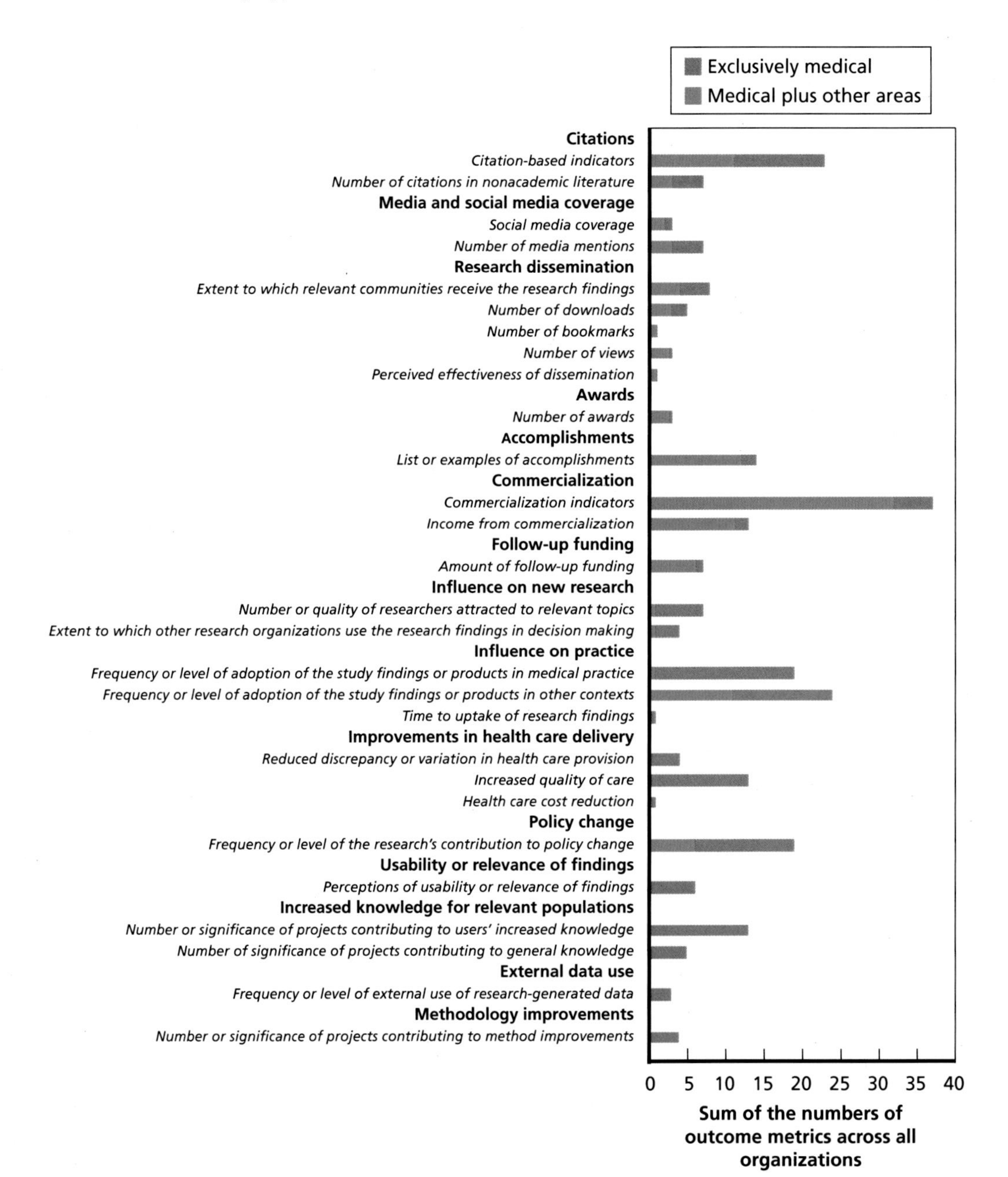

Figure C.10
Impact Metrics, by Type of Research

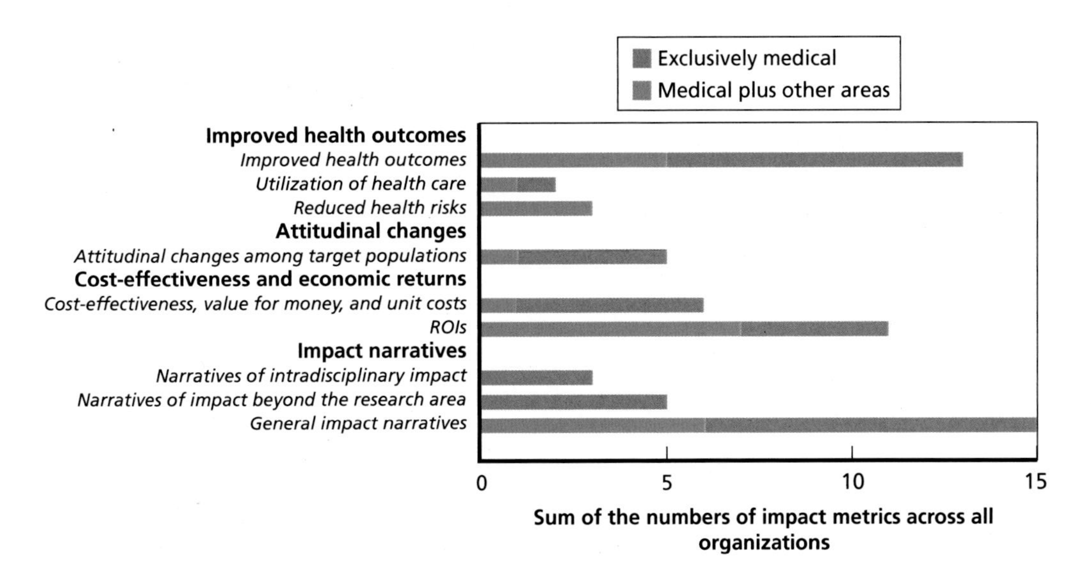

Frequency of Metric Categories, by Type of Research

Figure C.11
Input Metric Categories, by Type of Research

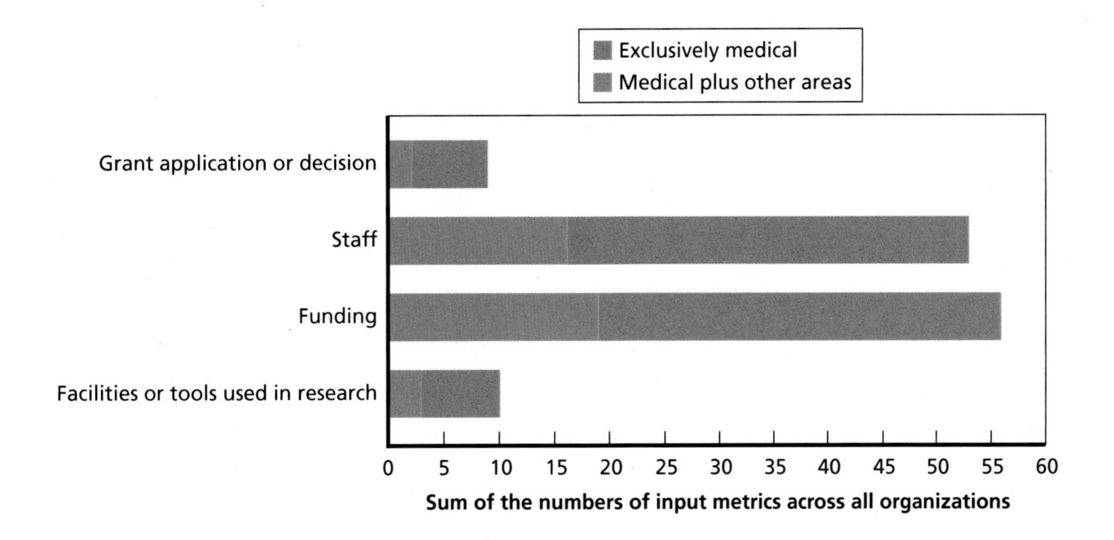

Figure C.12
Process Metric Categories, by Type of Research

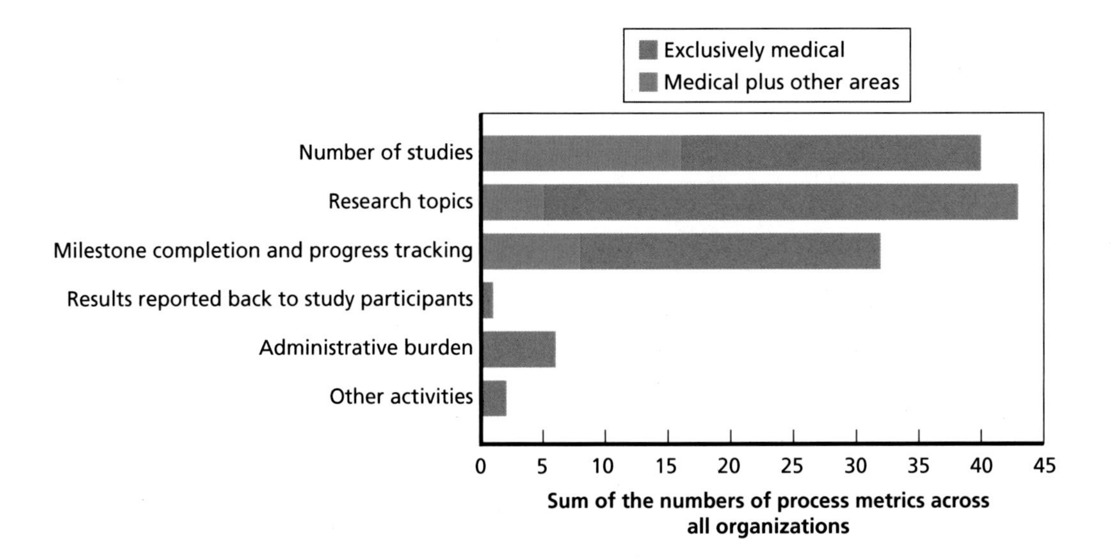

Figure C.13
Output Metric Categories, by Type of Research

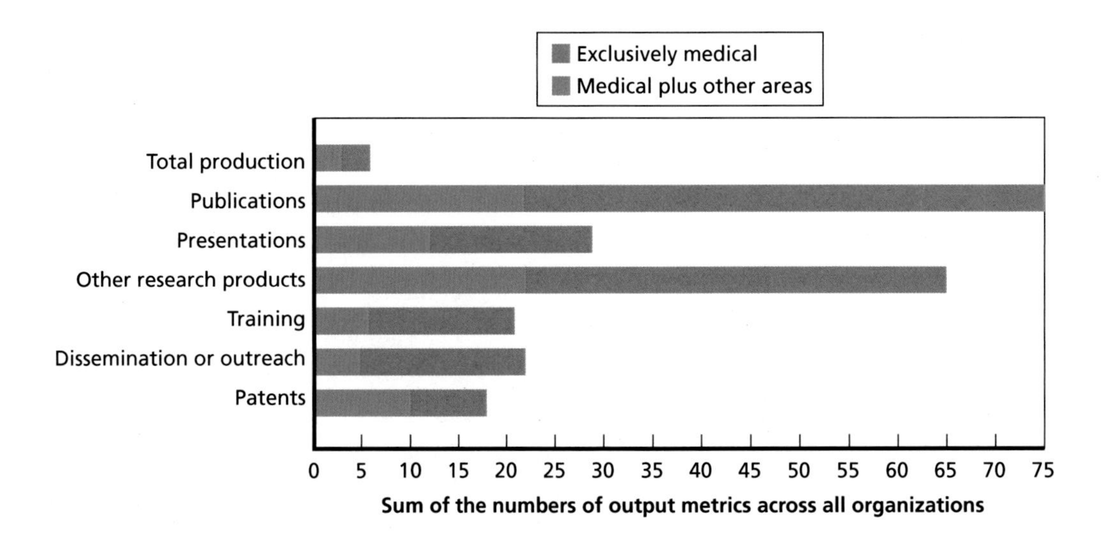

Figure C.14
Outcome Metric Categories, by Type of Research

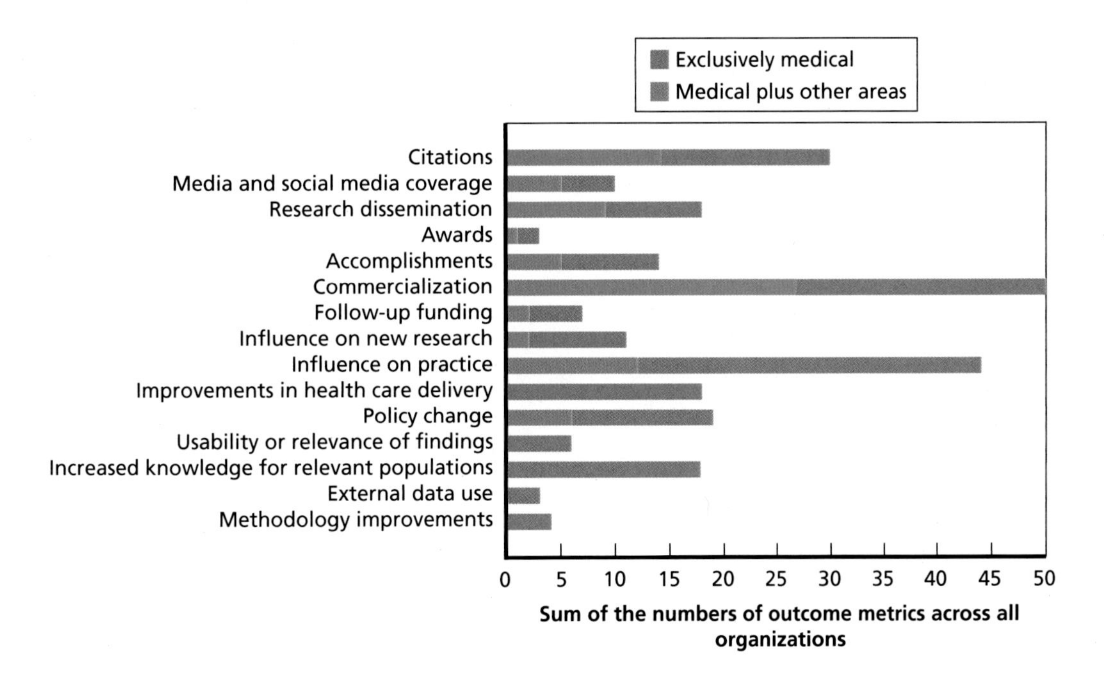

Figure C.15
Impact Metric Categories, by Type of Research

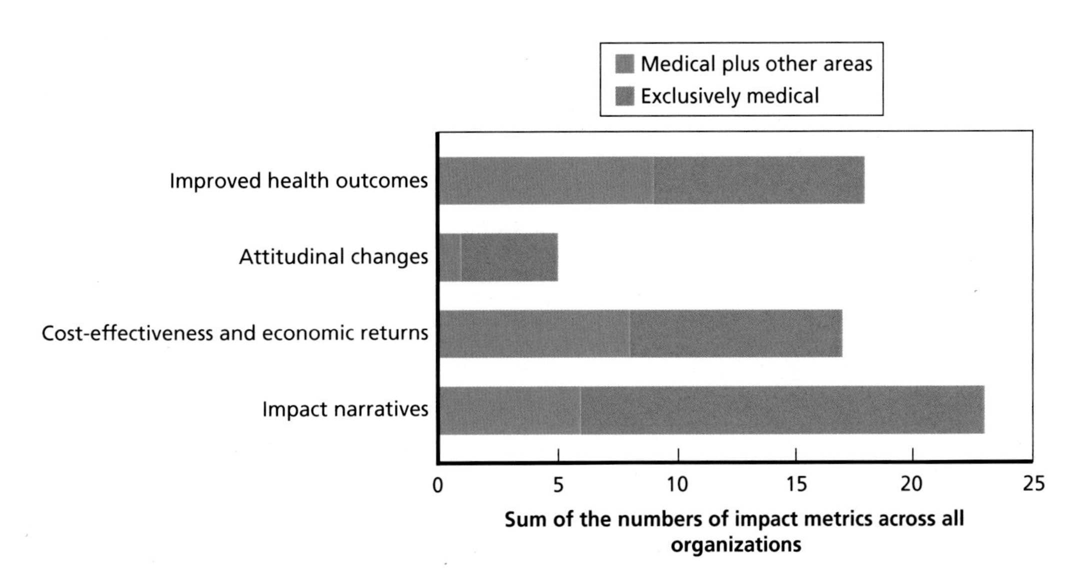

Frequency of Metric Types, by Type of Organization

Figure C.16
Input Metrics, by Type of Organization

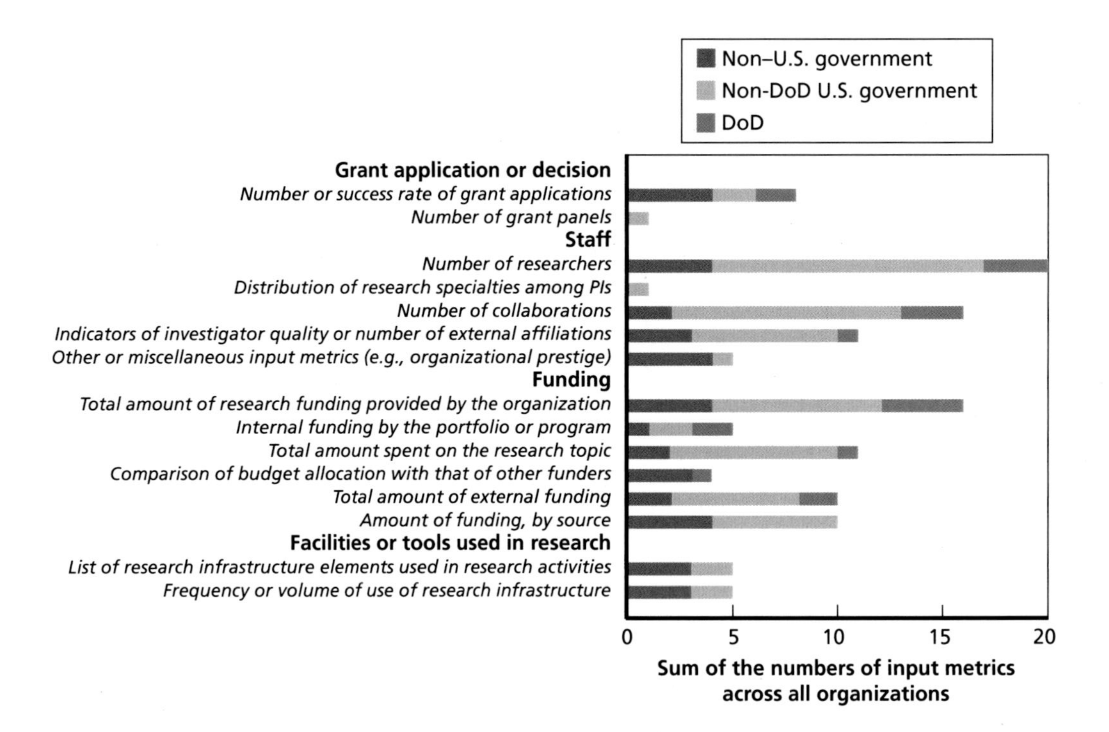

Figure C.17
Process Metrics, by Type of Organization

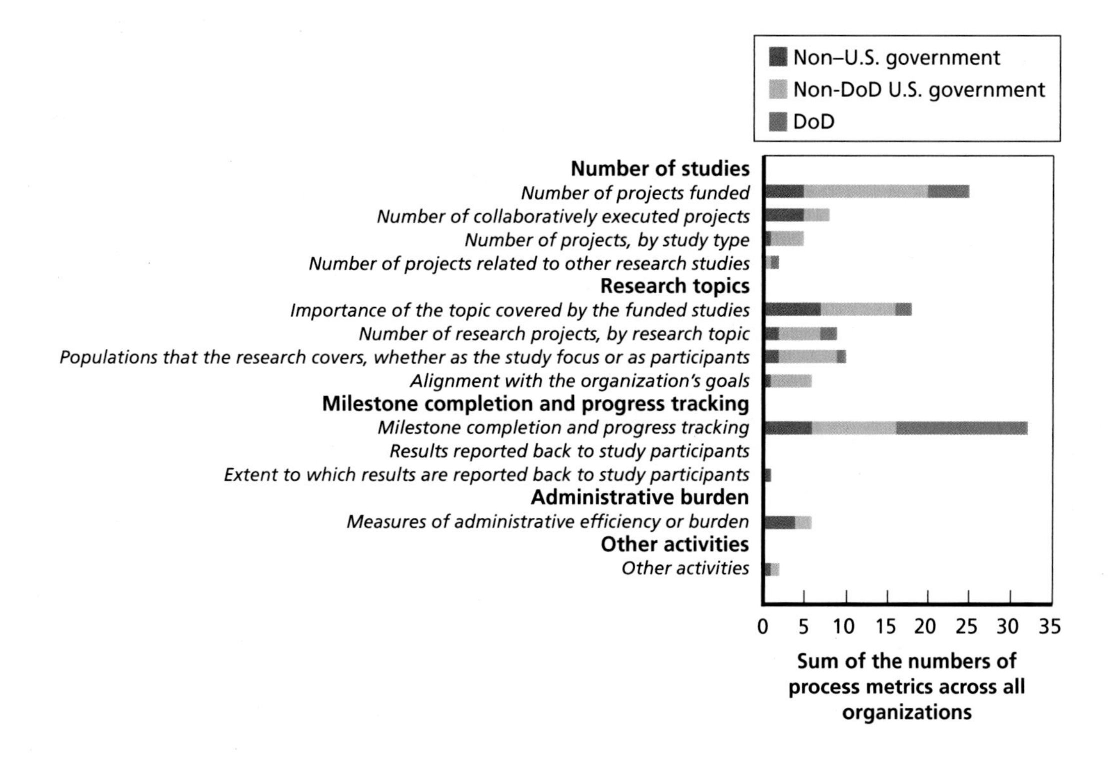

Figure C.18
Output Metrics, by Type of Organization

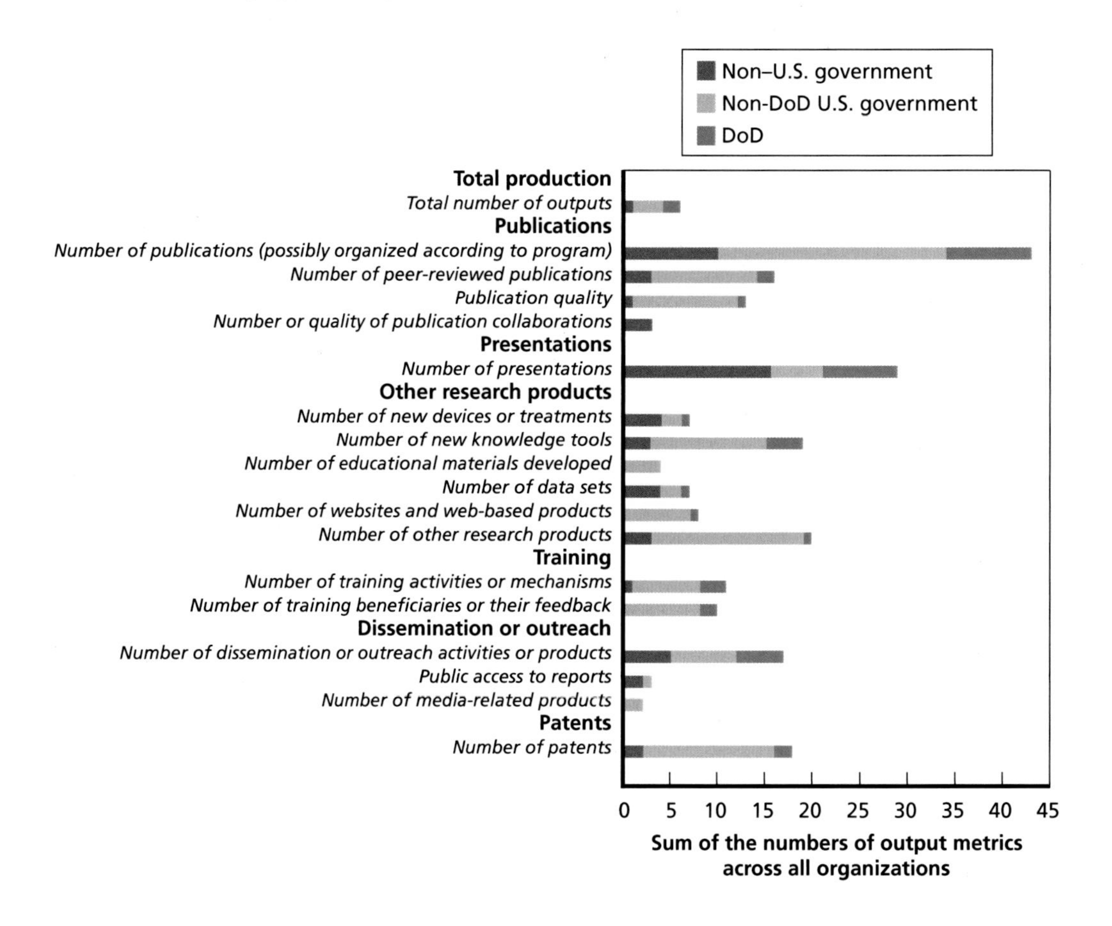

Figure C.19
Outcome Metrics, by Type of Organization

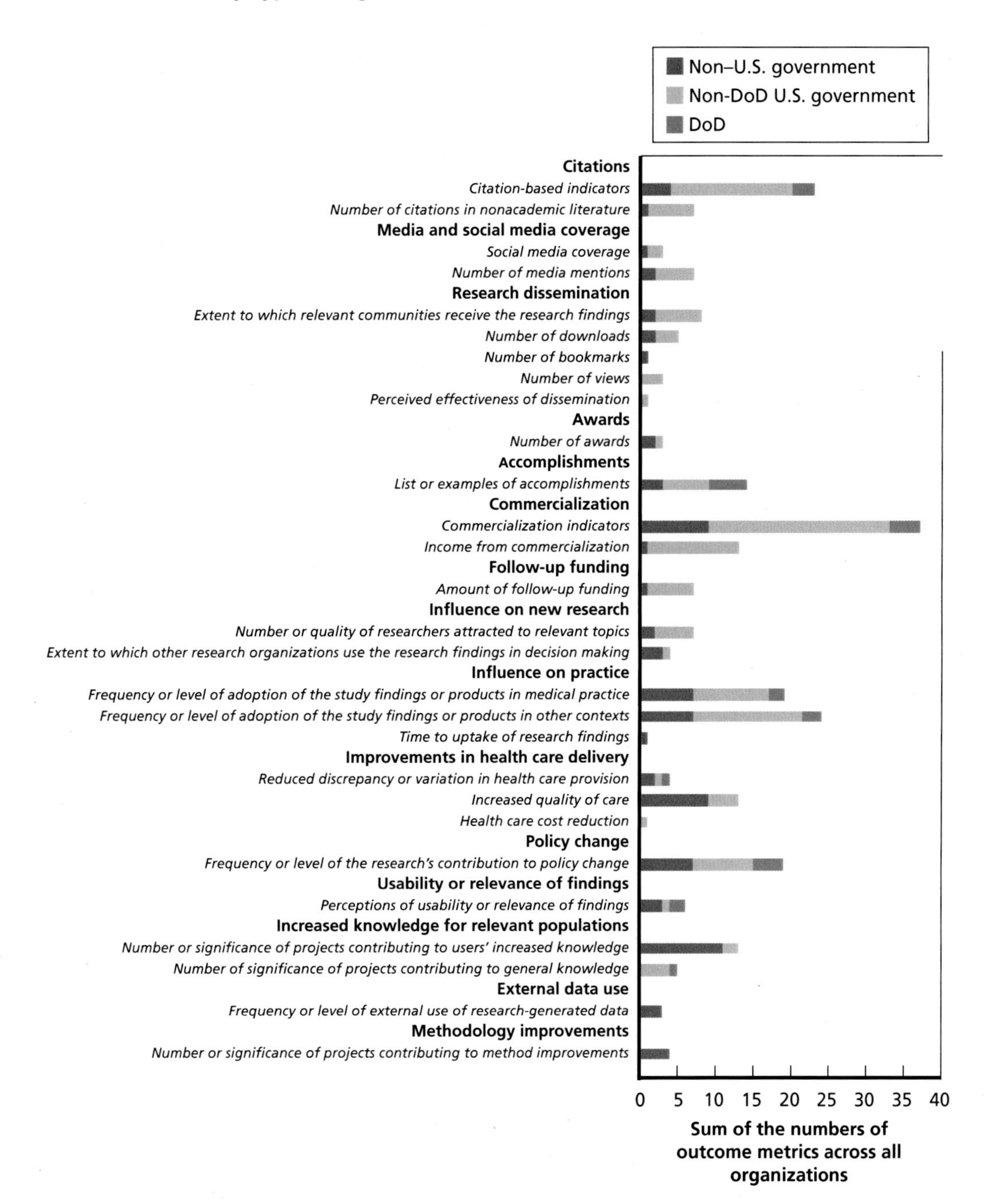

Figure C.20
Impact Metrics, by Type of Organization

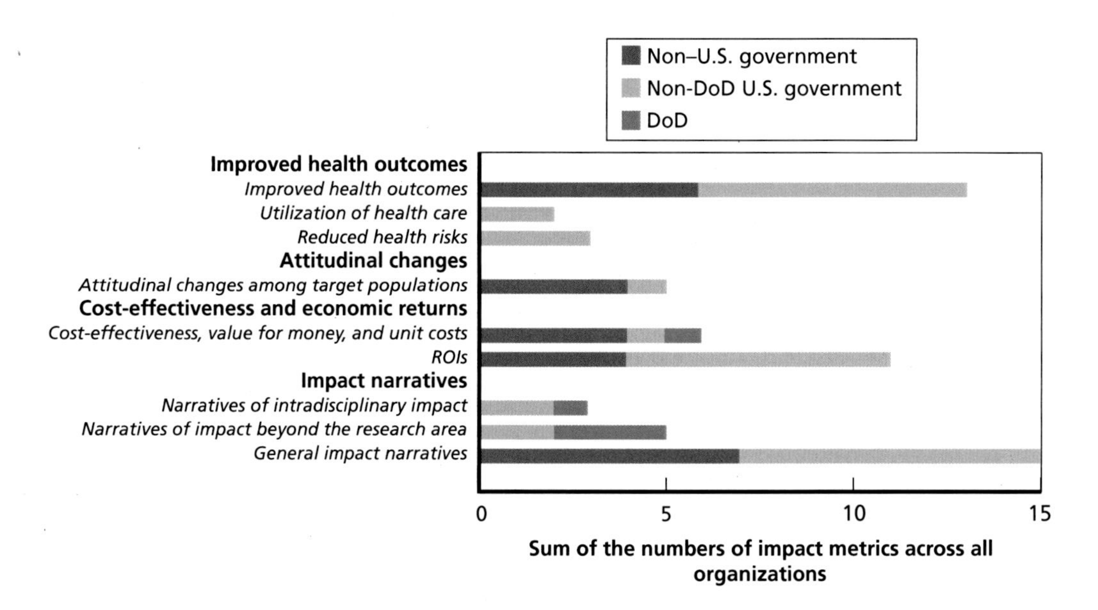

Frequency of Metric Categories, by Type of Organization

Figure C.21
Input Metric Categories, by Type of Organization

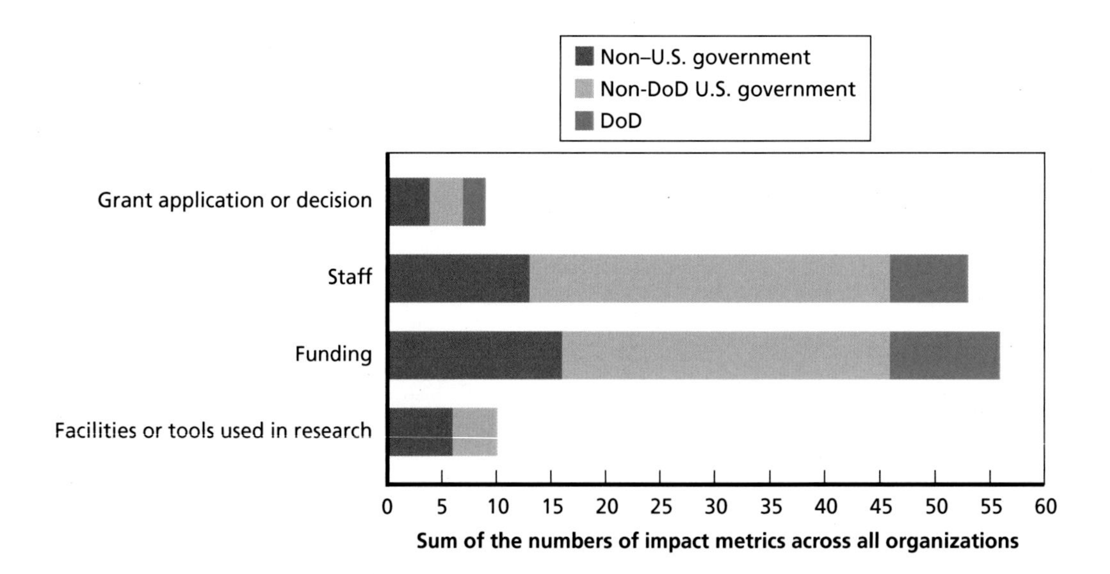

Figure C.22
Process Metric Categories, by Type of Organization

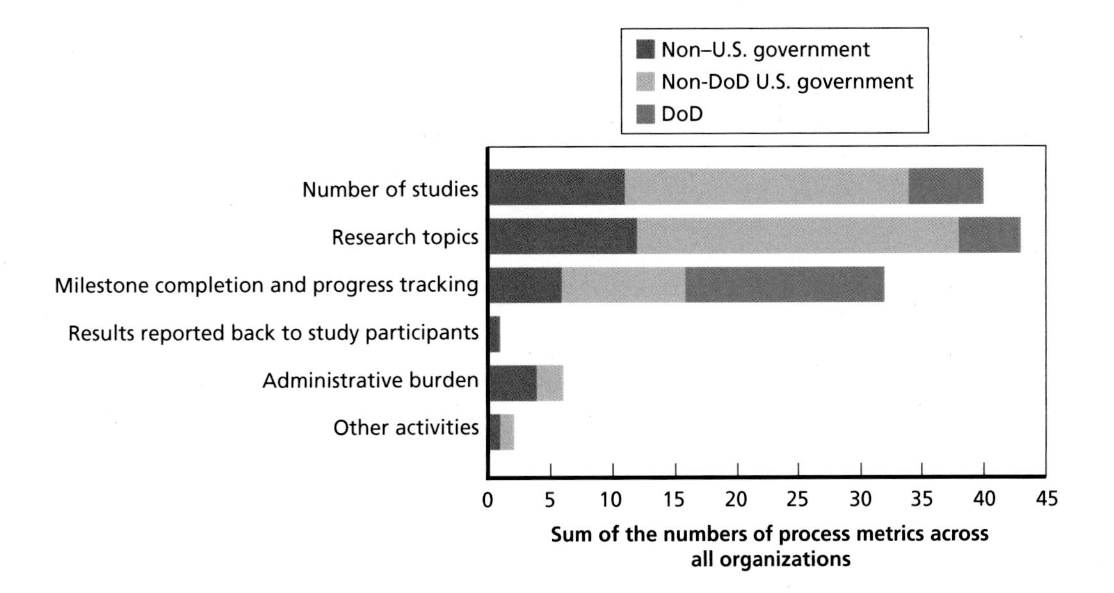

Figure C.23
Output Metric Categories, by Type of Organization

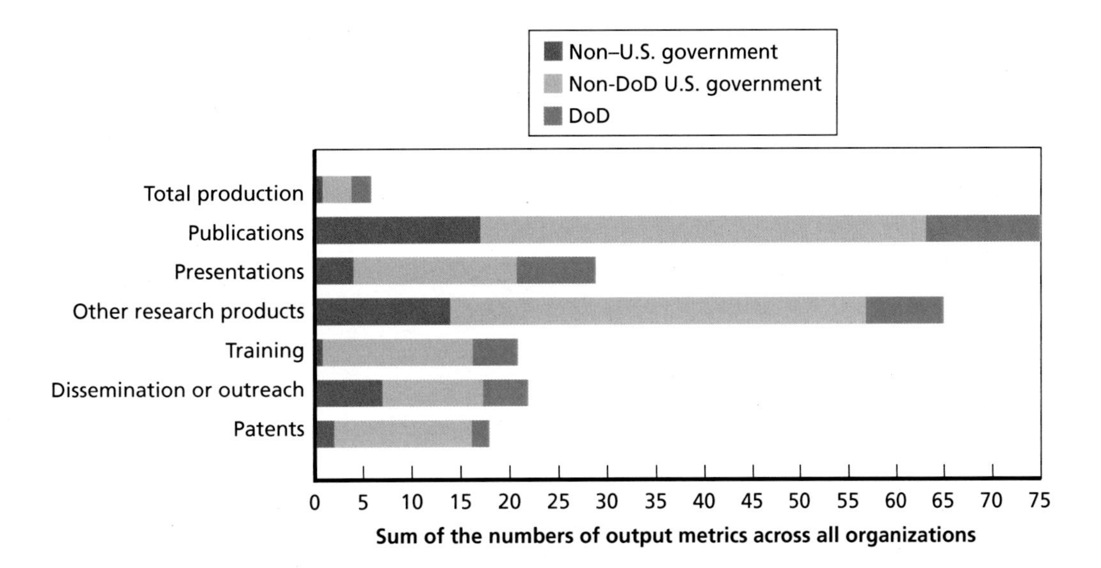

Figure C.24
Outcome Metric Categories, by Type of Organization

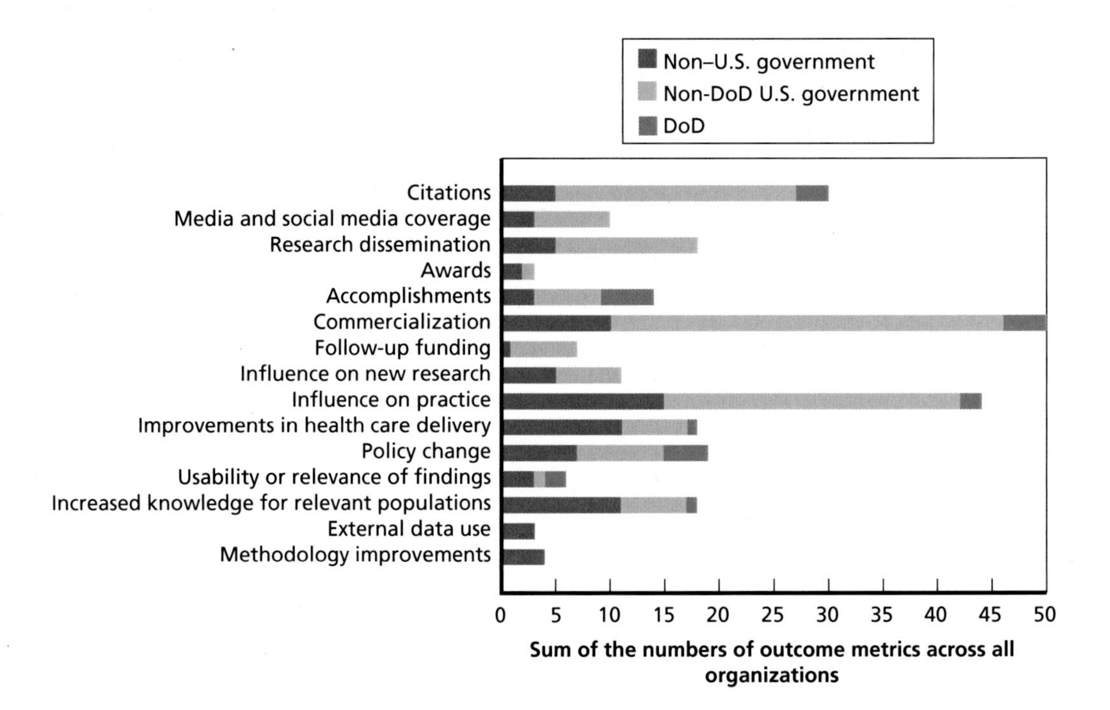

Figure C.25
Impact Metric Categories, by Type of Organization

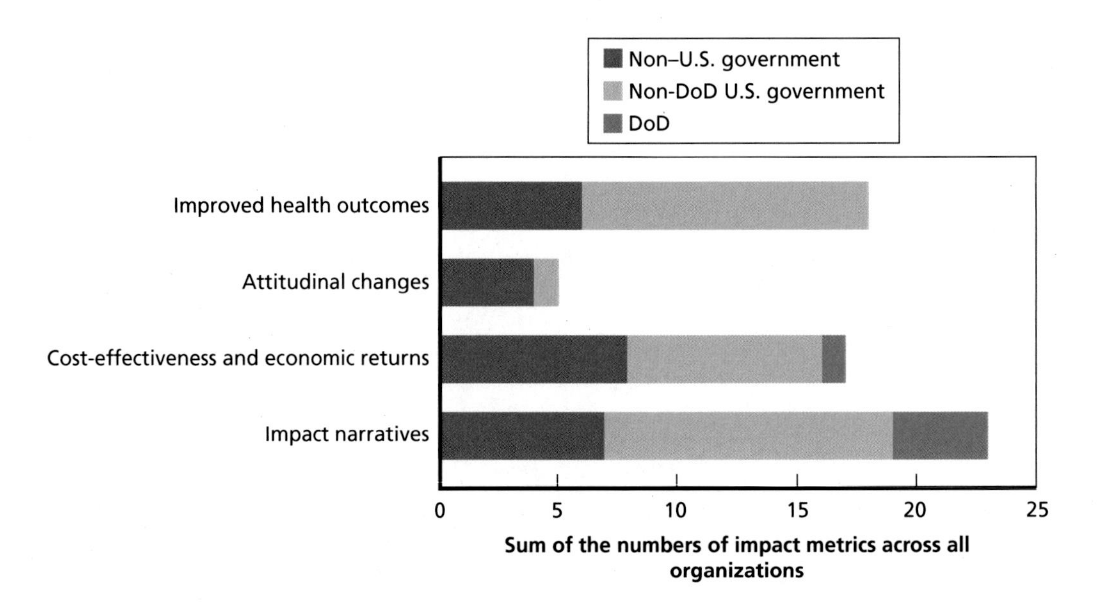

APPENDIX D

Stakeholder Interview Topic Guide

This appendix reproduces the topic guide we provided to stakeholders for each interview. We provide it here with only minor edits.

Introduction

At the request of the Defense Centers of Excellence for Psychological Health and Traumatic Brain Injury (DCoE), RAND has been asked to independently review methods of assessing research progress to inform performance-based decisions at the leadership level. This study is part of a multipronged effort associated to improve transparency, collaboration, and communication between DCoE programs and portfolios. This project aims to promote a better understanding of performance metrics applications and support compliance with DoD performance-reporting obligations.

One component of the study is a review of what portfolio performance metrics are used by other research organizations. As part of this effort, we are speaking with selected individuals from various organizations undertaking or funding research to learn about their approaches to assessing their research portfolios. We would like to speak with you today about these issues and benefit from your observations and insights.

Our working definition of a portfolio is the sum of research undertaken or funded by a given organization. This body of research consists of individual projects or studies. We appreciate [that] this conceptualization may not fit perfectly every context. For instance, some organizations may refer to their portfolios as programs. Therefore, we would be keen to hear what is your organization's understanding of what constitutes a research portfolio. Also, while we are primarily interested in metrics pertaining to research portfolios as a whole, we would be grateful for your perspective on all levels of analysis.

Our analysis focuses on the entire logic model underlying research evaluations, i.e., it examines components from inputs and activities to outputs, outcomes and impacts.

Please feel free to interrupt me with questions you might have or to seek clarification if a question is not clear.

Your participation in this interview is completely voluntary and the information you provide will be treated confidentially. We will not attribute comments to any individuals we have spoken to unless we have received an express permission to do so. We encourage you to be as candid as possible to ensure we develop findings that contribute to robust suggestions for improvement.

With your permission, we would like to make a recording of our conversation. This is for our internal purposes only; it will make it easier for us to go back to our notes to make sure we have not missed anything. We will not share the recording outside the project team and will destroy the recording at the completion of the project. Do we have your permission to record our discussion?

Key Questions (Tailored Based on Affiliation and Prior Document Review)

1. Could you please briefly describe the research portfolio your organization conducts or sponsors? In your view, what are the core objectives of the research portfolio of your organization? What constitutes a research portfolio in your organization?
   [Prompt: If your organization also organizes its research in programs, how do those differ from portfolios?]

2. What, if any, are your organization's responsibilities and existing processes with respect to measuring the performance of its research portfolio? Who is the audience for these assessments/who are these assessments reported to?
   [If not done yet, clarify what the interviewee means by "portfolio."]

   [Prompts: What is the objective of the performance measurement? What stage of the research process does it capture? Does it focus on research that is ongoing? On completed research? On both?]

3. What are the metrics your organization uses to assess the performance of its research portfolio? Could you please describe those?
   [Remember to ask about all components of the logic model if some not volunteered by the interviewee. Also reiterate if needed that this question asks about assessing portfolios as a whole.]

   [Prompts: How are the metrics selected? What are the limitations of these metrics? How frequently are the measurements undertaken? At what stage of the research project do these metrics apply/get collected? Whose responsibility is it to undertake the measurement?]

4. Does your organization have metrics to assess the performance of its individual research projects? Could you please describe those?
   [Prompt: How, if at all, are these metrics related to those used to assess the research portfolio as a whole?]

5. Are the metrics used by your organization summarized in a publicly available document? Is there any other documentation relevant to your research performance measurement you would be able to share with us?
   [Clarify: We are not interested in the actual values, we would appreciate even a redacted version.]

6. Does your organization coordinate its approach to performance measurement with other entities? If so, could you please describe how?

7. How well, in your opinion, does your organization's approach to research-portfolio measurement work?
[Prompts: What are the biggest obstacles to measuring research performance in your organization? What kind of performance is easiest to measure? What are the limitations of the metrics currently collected by your organization? Are there metrics you are not currently capturing but would like to? Why?]

8. Are there any other individuals you would recommend we consult with?

9. Is there anything else about research-portfolio metrics that you think we should know? About metrics for research projects? Are there any other issues you would like to comment on or discuss?

Thank you very much for your time and input.

APPENDIX E

Stakeholders Consulted for the Study

We would like to acknowledge the following people who agreed to be interviewed by the research team to inform this research (see Table E.1). In addition, four people preferred to remain anonymous. We are grateful to all our interviewees for sharing their valuable insights with the research team.

Table E.1
Stakeholders Consulted and Their Affiliations

Stakeholder	Affiliation
Sharon B. Arnold	AHRQ
Felix Barker	Vision CoE
Phillip Beatty	NIDILRR
David Bochner	NIH
Ruth Brannon	NIDILRR
Nigel Bush	T2
Marji Campbell	Deployment Health Clinical Center
Amanda Cash	HHS
Thomas J. Chapel	CDC
Joseph Cohn	DHA J-9
Clay Cooksey	CDC
Anand Desai	NSF
Sara Dodson	NIH
Rebecca Fisher	CDMRP
Darrin Frye	JPC-1
Tanisha Hammill	Hearing CoE
Lakeisha R. Henry	Hearing CoE
Lynn W. Henselman	Hearing CoE
Ron Hoover	Joint Program Committee 5 (Military Operational Medicine)
Amy M. Kilbourne	VA
Andrea Landes	Commonwealth Fund

Table E.1—Continued

Stakeholder	Affiliation
Melanie Lawson	CDC
Stanley S. Litow	IBM International Foundation
Daniel Lockney	NASA
Saafan Malik	DVBIC
Robert Mazzoli	Vision CoE
Emily Novicki	NIOSH
Cheryll Quirin	JPC-1
Rasheda Parks	NIDILRR
Philip Perlman	Howard Hughes Medical Institute
Amy Ratcliffe	PSI
Loretta Schlachta-Fairchild	JPC-1
David Thompson	JPC-1
Jude Tomasello	Joint Project Manager Medical Modeling and Simulation
Marina Volkov	NIH
Michael D. Walsh	NIST
Alice Y. Welch	FDA
Paul R. Zielinski	NIST

NOTE: T2 = National Center for Telehealth and Technology. J-9 = Research and Development Directorate. JPC-1 = Joint Program Committee 1 (Medical Simulation and Information Sciences).

APPENDIX F

The Research Portfolio Management Data Dictionary of the Former Defense Centers of Excellence for Psychological Health and Traumatic Brain Injury

This appendix reproduces a data dictionary developed and implemented by RPM. The information presented reflects the status of the tool as shared with the research team at the beginning of this research project. Although the dissolution of DCoE renders some aspects of that system obsolete, consultations undertaken during this study suggest that the system continues to provide an analytically useful reference point for surviving components or successors, including the DHA J-9. We provide it here with only the minimal corrections shown.

Background

Historically, the DCoE Research Portfolio Management (RPM) tracking spreadsheet has collected information on each [center's] research portfolio. This spreadsheet has primarily collected qualitative data which [are] difficult to analyze and validate. In an effort to develop [an] RPM tracking system that maintains up-to-date and reliable data, facilitates DCoE's ability to describe [its] research portfolio, and provides the opportunity to calculate metrics [that] measure the performance of the portfolio, a revised RPM tracking spreadsheet is recommended, which includes the . . . variables [described in the remainder of this appendix].

Key

Time of Entry	Color
Start of study	
Updated [q]uarterly	
End of [s]tudy	

Administrative Data

Variables listed in this category will provide the RPM office with data that will be used to track DCoE research studies from an administrative perspective.

Variable Name	Definition	Data Elements
DCoE_Unique_Identifier	Identifier created by DCoE to track research conducted by Centers	# [In this appendix, this symbol indicates "number."]
Center_Unique_Identifier	Identifier created at the Center level to track research	#
DCoE_Component_Center	DCoE Center conducting the study	1. [Deployment Health Clinical Center] 2. T2 3. DVBIC
Grant_or_Contract_No	The grant or contract number associated with the research protocol	#
Study_Title	Full title of the study (should match what is on the IRB protocol)	Text
Short_Study_Title	Shortened version of the study title [that] is used to refer to the study	Text
Study_Objective	Short Summary of the study's objective	Text
POP_beginning	[Period-of-performance] start date (based on funding document)	mm/dd/yyyy
POP_end	[Period-of-performance] end date (based on funding document)	mm/dd/yyyy
Principal_Investigator	Study PI for the IRB of record	Text
POC	Government lead/[contracting officer's representative] associated with the study	Text
Study_Status	Categorize the current status of this study	1. Pending IRB approval 2. IRB approved (no data) 3. Active Data Collection 4. Primary Data Analysis 5. Manuscript Preparation & Secondary Data Analysis 6. Closed (Work Complete)
Extension	Has an extension been approved for this study?	1. No 2. Yes
Revised_Study_Date	What is the new end date for this study?	mm/dd/yyyy
Extension_Reason	What is the purpose of the extension? (Please choose the category that best fits)	1. Data collection 2. Preparation of study deliverables 3. Data analysis 4. Manuscript development 5. Not Applicable

Variable Name	Definition	Data Elements
Study_Closure_Expected	Is all work on this study expected (This is separate [from] IRB or funding agency closure—refers to whether all work is complete)	0. No to end this FY? 1. Yes 2. Already Ended, All work Complete 3. Already Ended, Manuscript(s) Pending
IRB_Approval	Has IRB approval been obtained for this study?	0. No 1. Yes 2. In Process 3. Not Applicable
OMB Approval	Has OMB approval been obtained for this study [which collects data from non-DoD civilians]?* * **Note:** OMB approval is required	0. No 1. Yes 2. In Process 3. Not Applicable when data is collected from non-DoD civilians
Report Control Symbol	Has OMB approval been obtained for this study [which collects data from federal personnel and DoD civilians?]* * **Note:** OMB approval is required when data is collected from government personnel and DoD civilians	0. No 1. Yes 2. In Process 3. Not Applicable

Research Portfolio Description

Variables listed in this category will provide the RPM office with data that will be used to describe DCoE's research portfolio.

Variable Name	Definition	Data Elements
Type_of_Study	Type of study design (Choose the type of study that best fits)	1. Experimental/Intervention 2. Observational (Qualitative) 3. Observational (Analytic) 4. [Retrospective]/Secondary Data Analysis 5. Review 6. Meta Analysis 7. Feasibility Study 8. Usability Study 9. Cost Study
Survey_Focus Group	Does this study involve a survey or focus group?	0. No 1. Yes
Knowledge_Translation_Initiative	Specify whether this is an identified Knowledge Translation initiative	0. No 1. Yes
Knowledge_Translation_Element	Choose the Knowledge Translation element with which this study most closely aligns.* *See Definition of Terms [at the end of this appendix.]	1. Element 1 (needs & gaps assessment) 2. Element 2 (strategic analysis) 3. Element 3 (solution material development) 4. Element 4 (dissemination) 5. Element 5 (implementation) 6. Not Applicable

Variable Name	Definition	Data Elements
Center_Cross_Collaboration	Does this study involve cross collaboration between DCoE Centers?	0. No 1. Yes
NRAP	Research category from the National Research Action Plan (NRAP) continuum of care with which this study most closely aligns.* *See **Definition of Terms** [at the end of this appendix.]	1. Foundational science 2. Epidemiology 3. Etiology 4. Prevention and screening 5. Treatment 6. Follow up Care 7. Services Research
NRAP_2	Secondary NRAP category—any other category(ies) with which this study aligns (if applicable)	1. Foundational science 2. Epidemiology 3. Etiology 4. Prevention and screening 5. Treatment 6. Follow-up Care 7. Services Research 8. Not Applicable
Telehealth_Study	Specify whether this study involves telehealth	0. No 1. Yes
Primary_Content	Primary topic area of research	1. PH 2. TBI 3. Both 4. Other
Primary_Content_Other	If other, describe area of content	Text
TBI_Severity	Specify the category of Traumatic Brain Injury (TBI) on which this study primarily focuses.	1. Mild 2. Moderate 3. Severe 4. Not Applicable
PH_Content1	Primary psychological health (PH) topic area of research	1. PTSD [post-traumatic stress disorder] 2. Depression 3. SUD [substance-use disorder] 4. Suicide 5. General Mental Health/ Comorbidity 5. Not Applicable
PH_Content1_Topic	If "General Mental Health/ Comorbidity," specify topic area	Text
PH_Content2	Secondary PH topic area of research	1. PTSD 2. Depression 3. SUD 4. Suicide 5. General Mental Health/ Comorbidity 6. Not Applicable
PH_Content2_Topic	If "General Mental Health/ Comorbidity," specify topic area	Text
PH_Content3	Tertiary PH topic area of research	1. PTSD 2. Depression 3. SUD 4. Suicide 5. General Mental Health/ Comorbidity 6. Not Applicable

Variable Name	Definition	Data Elements
PH_Content3_Topic	If "General Mental Health/ Comorbidity," specify topic area	Text
Primary_Population	Primary population of interest studied	1. Service Member 2. Reserve 3. National Guard 4. Veteran 5. Beneficiary 6. Civilian
Secondary_Population	Secondary population of interest studied	1. Service Member 2. Reserve 3. National Guard 4. Veteran 5. Beneficiary 6. Civilian 7. Not Applicable
Tertiary_Population	Tertiary population of interest studied	1. Service Member 2. Reserve 3. National Guard 4. Veteran 5. Beneficiary 6. Civilian 7. Not Applicable
Percent_SM	Percentage of Service Member participants in the study sample	%
Percent_Veteran	Percentage of Veteran participants in the study sample	%
Percent_Beneficiary	Percentage of Beneficiary participants in the study sample	%
Percent_Civilian	Percentage of Service Member participants in the study sample	%
Primary_Service	Primary military service of interest studied	1. Army 2. Navy 3. Air Force 4. Marine Corps 5. Coast Guard 6. All Services 7. Not Applicable
Secondary_Service	Secondary military service of interest studied	1. Army 2. Navy 3. Air Force 4. Marine Corps 5. Coast Guard 6. Not Applicable
Tertiary_Service	Tertiary military service of interest studied	1. Army 2. Navy 3. Air Force 4. Marine Corps 5. Coast Guard 6. Not Applicable
Percent_Army	Percentage of Army participants in the study sample	%
Percent_Navy	Percentage of Navy participants in the study sample	%

Variable Name	Definition	Data Elements
Percent_Air Force	Percentage of Air Force participants in the study sample	%
Percent_Marines	Percentage of Marine [Corps] participants in the study sample	%
Percent_Coast Guard	Percentage of Coast Guard participants in the study sample	%
Site_Type	Whether the study data collection (or access) is being conducted at one site or more than one site	1. Single Site 2. Multi Site 3. Not Applicable
Site_Names	Name of institution and location of each site where data are being collected (or accessed)	Text list of sites and cities/states
Other_Sites	List any other sites [that] are engaged in the research (e.g., oversight, analysis, tool development) and their roles	Text list of sites and roles
Primary_Outcome	Was the primary outcome for this study found to be negative or positive?	1. Negative 2. Positive 3. Unknown
Primary _Outcome_Text	List primary outcome for this study	Text
Secondary_Outcome	Was the secondary outcome for this study found to be negative or positive? (If more than two, list additional outcomes and whether [negative]/[positive] in comments section)	0. Negative 1. Positive 2. Unknown
Secondary_Outcome_Text	List secondary outcome(s) for this study	Text
Primary_Outcome_Effect_Size	What was the effect size for the primary outcome?	1. Effective (Moderate/Large) 2. Effective (Small) 3. Not Effective 4. Not Applicable
Secondary_Outcome_Effect_Size	What was the size for the secondary outcome?	1. Effective (Moderate/Large) 2. Effective (Small) 3. Not Effective 4. Not Applicable

Research Finances

Variables listed in this category will provide the RPM office with data that will be used to describe the finances that support DCoE research.

Variable Name	Definition	Data Elements
Internal_Funding	Does this study use internal (DCoE [headquarters] or Center) funds?	0. No 1. Yes
Type_of_Funding	Whether study funding is from a DoD or non-DoD entity	1. Non Internal DoD Funding 2. Non-DoD Funding 3. DCoE [headquarters] Funding 4. Center Internal Funding 5. Not Applicable
DoD_Funding_Source	List DoD funding source(s)	Funding agency or agencies (text)
Funding_Category	What type of funds are used for this study?* * **Note:** Internally-funded studies use [operation and maintenance] fund	1. [Research, development, test, and evaluation] 2. [Operation and maintenance] 3. Both 4. Other 5. Not Applicable
Funding_Category_Other	If "Other," specify type of funds used.	Text
DoD_NON_Funding_ Source	List non-DoD funding source(s)	Funding agency/agencies (text)
Funding_Vehicle	The type of funding vehicle providing funds for this study	1. Intramural grant 2. Extramural grant 3. Contract ([federally funded research and development center]) 4. Contract (Non [federally funded research and development center]) 5. Other 6. None 7. Not Applicable
Funding_Vehicle_Other	If "Other," specify funding vehicle	Text
Appropriation_Date	Date funds were appropriated/ executed—**For RAND studies, enter in RAND-only section below**	mm/dd/yyyy
Total_Research_Budget	List the total budget for this study (all years)	#
Awardee_Type	Whether this study is the prime or sub awardee	1. Prime 2. Subaward 3. Not Applicable
Total_DCoE_Budget	Total budget awarded to DCoE (or Center)—if applicable	#
Total_Amount_FY	Total budget for current FY (if applicable)	#
Total_DCoE_Amount_FY	Total DCoE/Center budget for current FY (if applicable)	#
Monies_Spent_Quarter	Study funds spent current quarter of FY	#

Variable Name	Definition	Data Elements
Monies_Spent_Cumulative	Cumulative study funds spent through current quarter (all years)	#
Follow_On_Money_Expected	Is there additional money expected for this study?	0. No 1. Yes
Follow_On_Money_Date	When were the additional funds executed? **(Non-RAND studies)**	mm/dd/yyyy
Follow_On_Money_Amount	What was the amount of the additional funds? **(Non-RAND studies)**	#
	RAND Studies only	
RAND_Year1_Execution_Date	Contract year 1 date of funds execution	mm/dd/yyyy
RAND_Year1_Total Budget	Contract year 1 total budget	#
RAND_Year1_Cumulative	Contract year 1 study cumulative expenditures	#
RAND_Year2_Execution_Date	Contract year 2 date of funds execution	mm/dd/yyyy
RAND_Year2_Total Budget	Contract year 2 total budget	#
RAND_Year2_Cumulative	Contract year 2 study cumulative expenditures	#
RAND_Year3_Execution_Date	Contract year 3 date of funds execution	mm/dd/yyyy
RAND_Year3_Total Budget	Contract year 3 total budget	#
RAND_Year3_Cumulative	Contract year 3 study cumulative expenditures	#
RAND_Year4_Execution_Date	Contract year 4 date of funds execution	mm/dd/yyyy
RAND_Year4_Total Budget	Contract year 4 total budget	#
RAND_Year4_Cumulative	Contract year 4 study cumulative expenditures	#
RAND_Year5_Execution_Date	Contract year 5 date of funds execution	mm/dd/yyyy
RAND_Year5_Total Budget	Contract year 5 total budget	#
RAND_Year5_Cumulative	Contract year 5 study cumulative expenditures	#

Research Agreements

Variables listed in this category will provide the RPM office with data that will be used to describe DCoE collaborations and completed agreements that support DCoE research.

Variable Name	Definition	Data Elements
Num_Agreements	Total number of agreements associated with this study	#
Type_of_Agreement	Type of agreement associated with this study* * **Note:** Other Cooperative Agreements may include the following: data acquisition/use agreements, interagency agreements, business associate agreements, etc.	1. [Technology transition agreement] 2. [Memorandum of understanding] 3. [Memorandum of agreement] 4. [Cooperative research and development agreement] 5. Other Cooperative Agreement 6. 1+ of the above 7. Not Applicable
Name_of_Agreement	List names of agreements associated with this study (include purpose and who is engaged in agreement)	Text list of agreements (or Not Applicable)
Num_Partner_Organizations	Total number of partnering organizations associated with this study	#
Type_of_Partner_Organizations1	Type of partnering organization associated with this study	1. DoD 2. VA 3. Other Non DoD Federal Agency 4. [Uniformed Services University of the Health Sciences] 5. Other University 6. Non Profit 7. Private Company 8. Not Applicable
Name_of_Partner_Organizations1	List name of partnering organization associated with this study (include role of partner)	Text Name of Partner Organization (or Not Applicable)
Type_of_Partner_Organization2	Type of partnering organization associated with this study	1. DoD 2. VA 3. Other Non DoD Federal Agency 4. [Uniformed Services University of the Health Sciences] 5. Other University 6. Non Profit 7. Private Company 8. Not Applicable
Name_of_Partner_Organization2	List name of partnering organization associated with this study (include role of partner)	Text Name of Partner Organization (or Not Applicable)
Name_of_Partner_Organization3	List names of any additional partnering organizations associated with this study (include role of partner)	Text Names of Remaining Partner Organizations (or Not Applicable)
Patents	Does this study have any patents associated with it?	0. No 1. Yes
Patent_Numbers	Provide any patent numbers associated with this study	#

Research Risks

Variables listed in this category will provide the RPM office with data that will be used to identify risks to ongoing DCoE research.

Variable Name	Definition	Data Elements
Research_Risk	This question refers to project risk—*not* risk as it pertains to patients as defined by human subject protection regulations (e.g., did this study have regulatory delays, recruiting problems, a large number of adverse events, or any other issue which could cause the study to not be completed according to the study protocol, [statement of work], or period of performance?)* * **Note:** Please include any issue as listed above or any others which DCoE leadership should have visibility of due to impact to the study.	0. No 1. Yes
Risk_Description	What is the nature of the risk?	Text description of risk
Risk_Probability	Rate the probability of risk to this study (1 = low probability/9 = high probability)	1. 1–3 (Low Probability) 2. 4–6 (Medium Probability) 3. 7–9 (High Probability) 4. Not Applicable
Risk_Impact	What level is the potential impact of the risk to this study?	1. Low 2. Medium 3. High 4. Not Applicable
Risk_Score	Probability*Impact	# (or Not Applicable)
Risk_Response	What is the plan to mitigate the risk for this study?	Text Description (or Not Applicable)

Literature Contributions

Variables listed in this category will provide the RPM office with data that will be used to identify contributions of DCoE research to the state of the science.

Variable Name	Definition	Data Elements
Publications_To_Date	Total number of peer-reviewed publications associated with this study to date	#
Name_Publications_ To_Date	List names of peer-reviewed publications associated with this study to date	Text
Presentations_To_Date	Number [of] presentations associated with this study to date	#

Variable Name	Definition	Data Elements
Name_Presentations_To_Date	List names of presentations associated with this study to date	Text
Reports_To_Date	Total number of reports associated with this study to date (e.g. white paper, web-based publication)	#
Name_Reports_To_Date	List names of reports associated with this study to date.	Text
Comments	Add any additional information that is relevant to understanding this study	Text

Definition of Terms:

Knowledge Translation Elements:

Element 1: Needs and Gaps Assessment—To systematically identify and prioritize psychological health and TBI needs and clinical practice gaps to provide recommendations for [knowledge-translation] initiatives and inform future research needs.

Element 2: Strategic Analysis/External Review of Scientific Evidence—To identify and select evidence-based knowledge solutions with the greatest potential for impact in the MHS.

Element 3: Solution Materials Development—To develop or refine tangible artifacts that support the translation of an approved knowledge solution.

Element 4: Dissemination—To create awareness, understanding and favorable opinions of the knowledge solution by disseminating solution materials to targeted audiences.

Element 5: Implementation—To develop and execute an implementation plan that promotes adoption of the knowledge solution along with behavioral and clinical changes associated with the knowledge solution application in clinical and operational settings.

The *National Research Action Plan (NRAP)* identifies a research framework within which studies can be organized along a progression.

Foundational Science—Basic discovery science

Epidemiology—Population-level (to include at-risk) descriptive and characterization in nature; the study of the distribution of associations between health related states

Etiology—Neurobiological mechanisms of the disease, to include possible causes of disorder

Prevention and Screening—Prevention intervention at different stages of illness; screening measures; assessment tools and measurement; training

Treatment—Aimed at symptom amelioration (includes psychotherapies and drugs) at different stages of illness including refractory, chronic, relapse, relapse prevention; address co-morbidities

Follow up Care—Aimed at understanding continued care after treatment

Services Research—Focused on system of care improvements and provider and non–healthcare provider

References

Abt Associates, *Evaluation of the Medical Research Program at the Department of Veterans Affairs*, Bethesda, Md., 2012.

altmetrics, "Tools," undated. As of March 30, 2018:
http://altmetrics.org/tools/

Balas, E. A., and S. A. Boren, "Managing Clinical Knowledge for Health Care Improvement," *Yearbook of Medical Informatics*, Vol. 9, No. 1, 2000, pp. 65–70.

Bozeman, Barry, and Juan Rogers, "Strategic Management of Government-Sponsored R&D Portfolios," *Environment and Planning C: Politics and Space*, Vol. 19, No. 3, 2001, pp. 413–442.

Braun, Virginia, and Victoria Clarke, "Using Thematic Analysis in Psychology," *Qualitative Research in Psychology*, Vol. 3, No. 2, 2006, pp. 77–101.

Brutscher, Philipp-Bastian, Steven Wooding, and Jonathan Grant, *Health Research Evaluation Frameworks: An International Comparison*, Santa Monica, Calif.: RAND Corporation, TR-629-DH, 2008. As of August 20, 2018:
https://www.rand.org/pubs/technical_reports/TR629.html

CAHS—*See* Canadian Academy of Health Sciences.

Callaway, Ewen, "Beat It, Impact Factor! Publishing Elite Turns Against Controversial Metric," *Nature*, Vol. 535, No. 7611, July 8, 2016, pp. 210–211.

Canadian Academy of Health Sciences, *Making an Impact: A Preferred Framework and Indicators to Measure Returns on Investment in Health Research—Report of the Panel on the Return on Investments in Health Research*, Ottawa, January 2009. As of March 30, 2018:
http://www.cahs-acss.ca/wp-content/uploads/2011/09/ROI_FullReport.pdf

CEP—*See* Commission on Evidence-Based Policymaking.

Chalmers, Iain, Michael B. Bracken, Ben Djulbegovic, Silvio Garattini, Jonathan Grant, A. Metin Gülmezoglu, David W. Howells, John P. A. Ioannidis, and Sandy Oliver, "How to Increase Value and Reduce Waste When Research Priorities Are Set," *Lancet*, Vol. 383, No. 9912, 2014, pp. 156–165.

Commission on Evidence-Based Policymaking, *The Promise of Evidence-Based Policymaking: Report of the Commission on Evidence-Based Policymaking*, Washington, D.C., September 2017. As of December 19, 2017:
https://cep.gov/content/dam/cep/report/cep-final-report.pdf

Congressional Budget Office, *Federal Support for Research and Development*, Washington, D.C., Publication 2927, June 2007. As of February 15, 2018:
https://www.cbo.gov/sites/default/files/cbofiles/ftpdocs/82xx/doc8221/06-18-research.pdf

Coryn, Chris L. S, Lindsay A. Noakes, Carl D. Westine, and Daniela C. Schröter, "A Systematic Review of Theory-Driven Evaluation Practice from 1990 to 2009," *American Journal of Evaluation*, Vol. 32, No. 2, 2011, pp. 199–226.

Defense Health Board, *Improving Defense Health Program Medical Research Processes*, Falls Church, Va., August 8, 2017. As of December 30, 2017:
http://www.dtic.mil/dtic/tr/fulltext/u2/1038653.pdf

F1000Prime, "What Is F1000Prime?" undated. As of March 30, 2018:
https://f1000.com/prime/about/whatis

FDA—*See* U.S. Food and Drug Administration.

Felknor, Sarah A., Donjanea Fletcher Williams, and Sidney C. Soderholm, *National Occupational Research Agenda: Second Decade in Review 2006–2016*, Cincinnati, Ohio: U.S. Department of Health and Human Services, Centers for Disease Control and Prevention, National Institute for Occupational Safety and Health, DHHS (NIOSH) Publication 2017-146, 2017. As of August 21, 2018:
https://www.cdc.gov/niosh/docs/2017-146/pdfs/2017-146.pdf?id=10.26616/NIOSHPUB2017146

Fereday, Jennifer, and Eimear Muir-Cochrane, "Demonstrating Rigor Using Thematic Analysis: A Hybrid Approach of Inductive and Deductive Coding and Theme Development," *International Journal of Qualitative Methods*, Vol. 5, No. 1, 2006, pp. 80–92.

Ganann, Rebecca, Donna Ciliska, and Helen Thomas, "Expediting Systematic Reviews: Methods and Implications of Rapid Reviews," *Implementation Science*, Vol. 5, No. 1, 2010, p. 56.

GAO—*See* U.S. Government Accountability Office.

Gilmour, John B., "Implementing OMB's Program Assessment Rating Tool (PART): Meeting the Challenges of Integrating Budget and Performance," *OECD Journal on Budgeting*, Vol. 7, No. 1, May 22, 2007, pp. 1–40.

Gore, Kristie L., K. C. Osilla, E. Hoch, S. Brooks Holliday, and B. A. Bicksler, *Cataloguing Ongoing Psychological Health Research in the Department of Defense: Knowledge Management Recommendations*, Santa Monica, Calif.: RAND Corporation, unpublished research.

Grant, Jonathan, *Impact and the Research Excellence Framework: New Challenges for Universities*, Santa Monica, Calif.: RAND Corporation, CP-661, 2012. As of August 21, 2018:
https://www.rand.org/pubs/corporate_pubs/CP661.html

Grant, Jonathan, Philipp-Bastian Brutscher, Susan Guthrie, Linda Butler, and Steven Wooding, *Capturing Research Impacts: A Review of International Practice*, Santa Monica, Calif.: RAND Corporation, DB-578-HEFCE, 2010. As of August 20, 2018:
https://www.rand.org/pubs/documented_briefings/DB578.html

Grant, Jonathan, Liz Green, and Barbara Mason, "Basic Research and Health: A Reassessment of the Scientific Basis for the Support of Biomedical Science," *Research Evaluation*, Vol. 12, No. 3, December 1, 2003, pp. 217–224.

Grant, Jonathan, and Steven Wooding, *In Search of the Holy Grail: Understanding Research Success*, Santa Monica, Calif.: RAND Corporation, OP-295-GBF, 2010. As of August 20, 2018:
https://www.rand.org/pubs/occasional_papers/OP295.html

Greenfield, Victoria, Valerie L. Williams, and Elisa Eiseman, *Using Logic Models for Strategic Planning and Evaluation: Application to the National Center for Injury Prevention and Control*, Santa Monica, Calif.: RAND Corporation, TR-370-NCIPC, 2006. As of August 21, 2018:
https://www.rand.org/pubs/technical_reports/TR370.html

Greenhalgh, Trisha, James Raftery, Steve Hanney, and Matthew Glover, "Research Impact: A Narrative Review," *BMC Medicine*, Vol. 14, No. 78, 2016, pp. 1–16.

Growth in Labour Markets in Low Income Countries, *Call for Proposals, Phase IV: Specifications*, version 1.0, January 8, 2017.

Guthrie, Susan, Ioana Ghiga, and Steven Wooding, "What Do We Know About Grant Peer Review in the Health Sciences? [Version 2; Referees: 2 Approved]," *F1000Research*, Vol. 6, March 27, 2018, pp. 1–23.

Guthrie, Susan, Benoit Guerin, Helen Wu, Sharif Ismail, and Steven Wooding, *Alternatives to Peer Review in Research Project Funding: 2013 Update*, Santa Monica, Calif.: RAND Corporation, RR-139-DH, 2013. As of August 21, 2018:
https://www.rand.org/pubs/research_reports/RR139.html

Guthrie, Susan, Alexandra Pollitt, Stephen Hanney, and Jonathan Grant, *Investigating Time Lags and Attribution in the Translation of Cancer Research: A Case Study Approach*, Santa Monica, Calif.: RAND Corporation, RR-627-WT, 2014. As of September 18, 2018:
https://www.rand.org/pubs/research_reports/RR627.html

Guthrie, Susan, Watu Wamae, Stephanie Diepeveen, Steven Wooding, and Jonathan Grant, *Measuring Research: A Guide to Research Evaluation Frameworks and Tools*, Santa Monica, Calif.: RAND Corporation, MG-1217-AAMC, 2013. As of August 21, 2018:
https://www.rand.org/pubs/monographs/MG1217.html

Hall, Bronwyn H., Jacques Mairesse, and Pierre Mohnen, "Measuring the Returns to R&D," in Bronwyn H. Hall and Nathan Rosenberg, eds., *Handbook of the Economics of Innovation*, Vol. 2, Amsterdam: Elsevier, 2010, pp. 1033–1082.

HHS—*See* U.S. Department of Health and Human Services.

Hicks, Diana, Paul Wouters, Ludo Waltman, Sarah de Rijcke, and Ismail Rafols, "Bibliometrics: The Leiden Manifesto for Research Metrics," *Nature*, Vol. 520, No. 7548, April 22, 2015, pp. 429–431. As of August 21, 2018:
https://www.nature.com/news/bibliometrics-the-leiden-manifesto-for-research-metrics-1.17351

Hinrichs, Saba, Erin Montague, and Jonathan Grant, *Researchfish: A Forward Look—Challenges and Opportunities for Using Researchfish to Support Research Assessment*, London: Policy Institute, King's College, November 2015. As of August 21, 2018:
https://www.kcl.ac.uk/sspp/policy-institute/publications/Researchfish%20A%20forward%20look.pdf

Hinrichs-Krapels, Saba, and Jonathan Grant, "Exploring the Effectiveness, Efficiency and Equity (3e's) of Research and Research Impact Assessment," *Palgrave Communications*, Vol. 2, Article 16090, 2016.

Hirsch, J. E., "An Index to Quantify an Individual's Scientific Research Output," *Proceedings of the National Academy of Sciences of the United States of America*, Vol. 102, No. 46, November 15, 2005, pp. 16569–16572.

Hutchins, B. Ian, Xin Yuan, James M. Anderson, and George M. Santangelo, "Relative Citation Ratio (RCR): A New Metric That Uses Citation Rates to Measure Influence at the Article Level," *PLoS Biology*, September 6, 2016.

Institute for Research on Innovation and Science, "About IRIS," undated. As of April 3, 2018:
http://iris.isr.umich.edu/about/

Institute of Medicine, NAS, and National Academy of Engineering—*See* Institute of Medicine, National Academy of Sciences, and National Academy of Engineering.

Institute of Medicine, National Academy of Sciences, and National Academy of Engineering, *Implementing the Government Performance and Results Act for Research: A Status Report*, Washington, D.C.: National Academies Press, 2001. As of August 21, 2018:
http://www.nap.edu/catalog/10106/implementing-the-government-performance-and-results-act-for-research-a

IRIS—*See* Institute for Research on Innovation and Science.

Jones, Molly Morgan, Catriona Manville, and Joanna Chataway, "Learning from the UK's Research Impact Assessment Exercise: A Case Study of a Retrospective Impact Assessment Exercise and Questions for the Future," *Journal of Technology Transfer*, 2017.

Khangura, Sara, Kristin J. Danko, Rob Cushman, Jeremy Grimshaw, and David Moher, "Evidence Summaries: The Evolution of a Rapid Review Approach," *Systematic Reviews*, Vol. 1, No. 10, February 10, 2012.

Khangura, Sara, Julie Polisena, Tammy J. Clifford, Kelly Farrah, and Chris Kamel, "Rapid Review: An Emerging Approach to Evidence Synthesis in Health Technology Assessment," *International Journal of Technology Assessment in Health Care*, Vol. 30, No. 1, January 2014, pp. 20–27.

Klavans, Richard, and Kevin W. Boyack, "Research Portfolio Analysis and Topic Prominence," *Journal of Infometrics*, Vol. 11, No. 4, November 2017, pp. 1158–1174.

Landree, Eric, Hirokazu Miyake, and Victoria Greenfield, *Nanomaterial Safety in the Workplace: Pilot Project for Assessing the Impact of the NIOSH Nanotechnology Research Center*, Santa Monica, Calif.: RAND Corporation, RR-1108-NIOSH, 2015. As of August 21, 2018:
https://www.rand.org/pubs/research_reports/RR1108.html

Lariviere, Vincent, Veronique Kiermer, Catriona J. MacCallum, Marcia McNutt, Mark Patterson, Bernd Pulverer, Sowmya Swaminathan, Stuart Taylor, and Stephen Curry, "A Simple Proposal for the Publication of Journal Citation Distributions," *BioRxiv*, 062109, September 11, 2016.

Ledford, Heidi, "Universities Struggle to Make Patents Pay," *Nature*, Vol. 501, No. 7468, September 24, 2013, pp. 471–472.

Leiden Manifesto for Research Metrics, homepage, undated. As of February 13, 2017:
http://www.leidenmanifesto.org

Li, Danielle, and Leila Agha, "Big Names or Big Ideas: Do Peer-Review Panels Select the Best Science Proposals?" *Science*, Vol. 348, No. 6233, April 24, 2015, pp. 434–438. As of August 21, 2018:
http://science.sciencemag.org/content/348/6233/434

Linton, Jonathan, and Nicholas Vonortas, "From Research Project to Research Portfolio: Meeting Scale and Complexity," *Foresight-Russia*, Vol. 9, No. 2, 2015, pp. 38–43.

Manville, Catriona, Molly Morgan Jones, Marie-Louise Henham, Sophie Castle-Clarke, Michael Frearson, Salil Gunashekar, and Jonathan Grant, *Preparing Impact Submissions for REF 2014: An Evaluation—Approach and Evidence*, Santa Monica, Calif.: RAND Corporation, RR-726-HEFCE, 2015. As of September 18, 2018:
https://www.rand.org/pubs/research_reports/RR726.html

Martín-Martín, Alberto, Enrique Orduna-Malea, Juan M. Ayllón, and Emilio Delgado López-Cózar, *The Counting House: Measuring Those Who Count—Presence of Bibliometrics, Scientometrics, Infometrics, Webometrics and Altmetrics in the Google Scholar Citations, ResearcherID, ResearchGate, Mendeley and Twitter*, EC3 Working Paper 21, 2016.

Mayne, John, "Addressing Attribution Through Contribution Analysis: Using Performance Measures Sensibly," *Canadian Journal of Program Evaluation*, Vol. 16, No. 1, 2001, pp. 1–24.

MHS—*See* Military Health System.

Milat, A. J., A. E. Bauman, and S. Redman, "A Narrative Review of Research Impact Assessment Models and Methods," *Health Research Policy and Systems*, Vol. 13, No. 18, 2015, p. 1.

Military Health System, "Defense Health Agency," undated. As of September 18, 2018:
https://www.health.mil/dha

———, "DHA Symposium Brings Together Minds to Get the Most Out of Research Dollars," October 17, 2017. As of February 16, 2018:
https://health.mil/News/Articles/2017/10/17/
DHA-symposium-brings-together-minds-to-get-the-most-out-of-research-dollars

Morgan Jones, Molly, and Jonathan Grant, "Making the Grade: Methodologies for Assessing and Evidencing Research Impact," in Andrew Dean, Michael Wykes, and Hilary Stevens, eds., *7 Essays on Impact*, Exeter, UK: Definitions, Evidence, and Structures to Capture Research Impact and Benefits Project Report for Jisc, University of Exeter, 2013, pp. 25–43. As of August 22, 2018:
https://www.exeter.ac.uk/media/universityofexeter/research/ourresearchexcellence/describeproject/pdfs/2013_06_04_7_Essays_on_Impact_FINAL.pdf

Morris, Zoë Slote, Steven Wooding, and Jonathan Grant, "The Answer Is 17 Years, What Is the Question: Understanding Time Lags in Translational Research," *Journal of the Royal Society of Medicine*, Vol. 104, No. 12, 2011, pp. 510–520.

Moynihan, Donald P., and Stéphane Lavertu, *Does Involvement in Performance Management Routines Encourage Performance Information Use? Evaluating GPRA and PART*, Madison, Wis.: Robert M. La Follette School of Public Affairs Working Paper 2011-017, October 5, 2011. As of December 19, 2017:
http://www.lafollette.wisc.edu/images/publications/workingpapers/moynihan2011-017.pdf

———, "Does Involvement in Performance Management Routines Encourage Performance Information Use? Evaluating GPRA and PART," *Public Administration Review*, Vol. 72, No. 4, July–August 2012, pp. 592–602.

Murphy, Kevin M., and Robert H. Topel, *Measuring the Gains from Medical Research: An Economic Approach*, Chicago: University of Chicago Press, 2003.

NASA—*See* National Aeronautics and Space Administration.

National Academies of Sciences, Engineering, and Medicine, *Evaluation of the Congressionally Directed Medical Research Programs Review Process*, Washington, D.C.: National Academies Press, 2016. As of August 21, 2018:
https://www.nap.edu/catalog/23652/
evaluation-of-the-congressionally-directed-medical-research-programs-review-process

———, *Principles and Practices for Federal Program Evaluation: Proceedings of a Workshop—in Brief*, Washington, D.C.: National Academies Press, March 2017a. As of August 21, 2018:
https://www.nap.edu/catalog/24716/principles-and-practices-for-federal-program-evaluation-proceedings-of-a

———, *Beyond Patents: Assessing the Value and Impact of Research Investments: Proceedings of a Workshop—in Brief*, Washington, D.C.: National Academies Press, October 2017b. As of August 21, 2018:
https://www.nap.edu/catalog/24920/beyond-patents-assessing-the-value-and-impact-of-research-investments

National Aeronautics and Space Administration, "NASA *Spinoff*," undated. As of October 12, 2018:
https://spinoff.nasa.gov/

National Aeronautics and Space Administration Technology Transfer Program, homepage, undated. As of September 17, 2018:
https://technology.nasa.gov/

National Institute of Standards and Technology, Technology Partnerships Office, "Federal Laboratory (Interagency) Technology Transfer Summary Reports," updated April 30, 2018a. As of February 14, 2018:
https://www.nist.gov/tpo/federal-laboratory-interagency-technology-transfer-summary-reports

———, "About the Technology Partnerships Office," updated June 25, 2018b. As of December 11, 2017:
https://www.nist.gov/tpo/about-technology-partnerships-office

National Institutes of Health, Office of Portfolio Analysis, last reviewed June 19, 2018. As of December 10, 2017:
https://dpcpsi.nih.gov/opa

National Institutes of Health Research Portfolio Online Reporting Tools, "NIH RePORTER Version 7.32.0," accessed February 12, 2018. As of October 15, 2018, current version:
https://projectreporter.nih.gov/reporter.cfm

National Research Council, *Furthering America's Research Enterprise*, Washington, D.C.: National Academies Press, 2014. As of August 21, 2018:
http://www.nap.edu/catalog/18804/furthering-americas-research-enterprise

National Science Foundation, Evaluation and Assessment Capability, "Evaluation Summary," undated (a). As of December 10, 2017:
https://www.nsf.gov/od/oia/eac/evaluation-inventory.jsp

———, *Format for Use in Submission of Interim and Final Research Performance Progress Reports*, Alexandria, Va., undated (b). As of December 10, 2017:
https://www.nsf.gov/bfa/dias/policy/rppr/frpprformat_2016.pdf

———, "Research Performance Progress Report (RPPR)," undated (c). As of December 10, 2017:
https://www.nsf.gov/bfa/dias/policy/rppr/

———, *FY 2017 NSF Budget Request to Congress: NSF Evaluation and Assessment Capability*, c. 2016. As of December 10, 2017:
https://www.nsf.gov/about/budget/fy2017/pdf/47_fy2017.pdf

———, Policy Office, Division of Institution and Award Support, "NSF Proposal Processing and Review," in *Proposal and Award Policies and Procedures Guide*, Alexandria, Va., NSF 17-1, effective January 30, 2017. As of September 18, 2018:
https://www.nsf.gov/pubs/policydocs/pappg17_1/nsf17_1.pdf

———, *Building the Future: Investing in Discovery and Innovation—NSF Strategic Plan for Fiscal Years (FY) 2018–2022*, Alexandria, Va., February 2018. As of February 14, 2018:
https://www.nsf.gov/pubs/2018/nsf18045/nsf18045.pdf

NIH—*See* National Institutes of Health.

NIH RePORTER—*See* National Institutes of Health Research Portfolio Online Reporting Tools.
NIST—*See* National Institute of Standards and Technology.

NSF—*See* National Science Foundation.

OECD—*See* Organisation for Economic Co-operation and Development.

Office of Management and Budget, "Commission on Evidence Based Policymaking," Washington, D.C., c. 2016. As of December 19, 2017:
https://obamawhitehouse.archives.gov/omb/management/commission_evidence

———, *Preparation, Submission, and Execution of the Budget*, Washington, D.C., Circular A-11, July 2017. As of February 14, 2018:
https://www.whitehouse.gov/sites/whitehouse.gov/files/omb/assets/a11_current_year/a11_2017.pdf

Office of Science and Technology Policy, "NSTC," undated. As of February 13, 2018:
https://www.whitehouse.gov/ostp/nstc/

OMB—*See* Office of Management and Budget.

Organisation for Economic Co-operation and Development, *Enhancing Research Performance Through Evaluation, Impact Assessment and Priority Setting*, Directorate for Science, Technology and Innovation, c. 2009. As of December 10, 2017:
http://www.oecd.org/sti/inno/Enhancing-Public-Research-Performance.pdf

Ovseiko, Pavel V., Trisha Greenhalgh, Paula Adam, Jonathan Grant, Saba Hinrichs-Krapels, Kathryn E. Graham, Pamela A. Valentine, Omar Sued, Omar F. Boukhris, Nada M. Al Olaqi, Idrees S. Al Rahbi, Anne-Maree Dowd, Sara Bice, Tamika L. Heiden, Michael D. Fischer, Sue Dopson, Robyn Norton, Alexandra Pollitt, Steven Wooding, Gert V. Balling, Ulla Jakobsen, Ellen Kuhlmann, Ineke Klinge, Linda H. Pololi, Reshma Jagsi, Helen Lawton Smith, Henry Etzkowitz, Mathias W. Nielsen, Carme Carrion, Maite Solans-Domènech, Esther Vizcaino, Lin Naing, Quentin H. N. Cheok, Baerbel Eckelmann, Moses C. Simuyemba, Temwa Msiska, Giovanna Declich, Laurel D. Edmunds, Vasiliki Kiparoglou, Alison M. J. Buchan, Catherine Williamson, Graham M. Lord, Keith M. Channon, Rebecca Surender, and Alastair M. Buchan, "A Global Call for Action to Include Gender in Research Impact Assessment," *Health Research Policy and Systems*, Vol. 14, No. 50, 2016, pp. 1–12.

Paul, Christopher, Jessica Yeats, Colin P. Clarke, and Miriam Matthews, *Assessing and Evaluating Department of Defense Efforts to Inform, Influence, and Persuade: Desk Reference*, Santa Monica, Calif.: RAND Corporation, RR-809/1-OSD, 2015. As of August 21, 2018:
https://www.rand.org/pubs/research_reports/RR809z1.html

Paul, Christopher, Jessica Yeats, Colin P. Clarke, Miriam Matthews, and Lauren Skrabala, *Assessing and Evaluating Department of Defense Efforts to Inform, Influence, and Persuade: Handbook for Practitioners*, Santa Monica, Calif.: RAND Corporation, RR-809/2-OSD, 2015. As of August 21, 2018:
https://www.rand.org/pubs/research_reports/RR809z2.html

Penfield, Teresa, Matthew J. Baker, Rosa Scoble, and Michael C. Wykes, "Assessment, Evaluations, and Definitions of Research Impact: A Review," *Research Evaluation*, Vol. 23, No. 1, January 1, 2014, pp. 21–32.

Population Services International, "PSI at a Glance," undated. As of December 20, 2017:
http://www.psi.org/about/at-a-glance/

President's Management Council and Executive Office of the President, *President's Management Agenda*, Washington, D.C., March 2018. As of September 19, 2018:
https://www.whitehouse.gov/wp-content/uploads/2018/03/
The-President%E2%80%99s-Management-Agenda.pdf

PSI—*See* Population Services International.

Public Law 103-62, Government Performance and Results Act of 1993, August 3, 1993.

Public Law 111-352, GPRA Modernization Act of 2010, January 4, 2011. As of August 22, 2018:
https://www.gpo.gov/fdsys/pkg/PLAW-111publ352/content-detail.html

Rafols, Ismael, and Alfredo Yegros, *Is Research Responding to Health Needs?* SSRN Papers, posted January 30, 2018; last revised March 26, 2018. As of February 13, 2018:
https://papers.ssrn.com/sol3/papers.cfm?abstract_id=3106713

REF—*See* Research Excellence Framework.

Research Excellence Framework, homepage, undated (a). As of October 6, 2019:
http://www.ref.ac.uk/

———, "Guidance," undated (b). As of October 6, 2019:
http://www.ref.ac.uk/guidance/

———, *Assessment Framework and Guidance on Submissions*, REF 02.2011, July 2011, updated to include addendum published in January 2012, January 2012. As of March 30, 2018:
http://www.ref.ac.uk/2014/media/ref/content/pub/assessmentframeworkandguidanceonsubmissions/GOS%20including%20addendum.pdf

Researchfish, "Our Members," undated. As of December 10, 2017:
https://www.researchfish.net/members

Ruegg, Rosalie T., "Quantitative Portfolio Evaluation of US Federal Research and Development Programs," *Science and Public Policy*, Vol. 34, No. 10, December 1, 2007, pp. 723–730.

Sampat, Bhaven, and Heidi L. Williams, *How Do Patents Affect Follow-On Innovation? Evidence from the Human Genome*, Cambridge, Mass.: National Bureau of Economic Research Working Paper 21666, 2017. As of April 1, 2018:
http://www.nber.org/papers/w21666

Savitz, Scott, Miriam Matthews, and Sarah Weilant, *Assessing Impact to Inform Decisions: A Toolkit on Measures for Policymakers*, Santa Monica, Calif.: RAND Corporation, TL-263-OSD, 2017. As of August 21, 2018:
https://www.rand.org/pubs/tools/TL263.html

Savitz, Scott, Henry H. Willis, Aaron C. Davenport, Martina Melliand, William Sasser, Elizabeth Tencza, and Dulani Woods, *Enhancing U.S. Coast Guard Metrics*, Santa Monica, Calif.: RAND Corporation, RR-1173-USCG, 2015. As of September 12, 2017:
https://www.rand.org/pubs/research_reports/RR1173.html

Shoemaker, Sarah J., Harmon Jordan, Kimberlee Luc, Yann Kumin, Amy Fitzpatrick, Kimberly Fredericks, and Sheila Weiss, *Evaluation of AHRQ's Pharmaceutical Outcomes Portfolio: Final Report*, Rockville, Md.: Agency for Healthcare Research and Quality, December 2007. As of August 21, 2018:
https://archive.ahrq.gov/research/findings/final-reports/pharmportfolio/index.html

Srivastava, Christina Viola, Nathaniel Deshmukh Towery, and Brian Zuckerman, "Challenges and Opportunities for Research Portfolio Analysis, Management, and Evaluation," *Research Evaluation*, Vol. 16, No. 3, September 2007, pp. 152–156.

STAR METRICS, "News: Announcement on Level I Activities," May 4, 2015. As of March 29, 2018:
https://www.starmetrics.nih.gov/Star/News

———, "Federal RePORTER," accessed February 12, 2018. As of August 25, 2018, current version:
https://federalreporter.nih.gov

Sugimoto, Cassidy R., Sam Work, Vincent Larivière, and Stefanie Haustein, "Scholarly Use of Social Media and Altmetrics: A Review of the Literature," *Journal of the Association for Information Science and Technology*, Vol. 68, No. 9, September 2017, pp. 2037–2062.

Taylor-Powell, Ellen, and Ellen Henert, *Developing a Logic Model: Teaching and Training Guide*, University of Wisconsin—Extension, Cooperative Extension Program Development and Evaluation, February 2008. As of February 14, 2018:
https://fyi.uwex.edu/programdevelopment/files/2016/03/lmguidecomplete.pdf

Tricco, Andrea C., Jesmin Antony, Wasifa Zarin, Lisa Strifler, Marco Ghassemi, John Ivory, Laure Perrier, Brian Hutton, David Moher, and Sharon E. Straus, "A Scoping Review of Rapid Review Methods," *BMC Medicine*, Vol. 13, No. 224, 2015, pp. 1–15.

ÜberResearch, "Decision Support Systems for Science Funders," undated. As of March 30: https://www.uberresearch.com/

U.S. Department of Agriculture, Agricultural Research Service, Office of Technology Transfer, "Office of Technology Transfer," undated. As of December 11, 2017: https://www.ars.usda.gov/office-of-technology-transfer/

U.S. Department of Health and Human Services, Office of the Assistant Secretary for Planning and Evaluation, "HHS Data Council: Introduction," undated. As of December 19, 2017: https://aspe.hhs.gov/hhs-data-council-introduction

———, "Strategic Plan FY 2014–2018," content last reviewed February 28, 2018. As of February 14, 2018: https://www.hhs.gov/about/strategic-plan/index.html

U.S. Food and Drug Administration, "FDA Technology Transfer Program," last updated August 8, 2018. As of December 11, 2017: https://www.fda.gov/scienceresearch/collaborativeopportunities/

U.S. Government Accountability Office, *Managing for Results: Greater Transparency Needed in Public Reporting on the Quality of Performance Information for Selected Agencies' Priority Goals*, Washington, D.C., GAO-15-788, September 10, 2015a. As of December 10, 2017: http://www.gao.gov/products/GAO-15-788

———, *Managing for Results: Implementation of GPRA Modernization Act Has Yielded Mixed Progress in Addressing Pressing Governance Challenges*, Washington, D.C., GAO-15-819, September 30, 2015b. As of December 10, 2017: http://www.gao.gov/products/GAO-15-819

———, *Managing for Results: Agencies Need to Fully Identify and Report Major Management Challenges and Actions to Resolve Them in Their Agency Performance Plans*, Washington, D.C., GAO-16-510, June 15, 2016. As of December 10, 2017: http://www.gao.gov/products/GAO-16-510

———, *Managing for Results: Government-Wide Actions Needed to Improve Agencies' Use of Performance Information in Decision Making*, Washington, D.C., GAO-18-609SP, September 5, 2018. As of September 19, 2018: https://www.gao.gov/products/GAO-18-609SP

Wallace, Matthew L., and Ismael Rafols, "Research Portfolio Analysis in Science Policy: Moving from Financial Returns to Societal Benefits," *Minerva*, Vol. 53, No. 2, June 2015, pp. 89–115.

Williams, Valerie L., Elisa Eiseman, Eric Landree, and David M. Adamson, *Demonstrating and Communicating Research Impact: Preparing NIOSH Programs for External Review*, Santa Monica, Calif.: RAND Corporation, MG-809-NIOSH, 2009. As of August 21, 2018: https://www.rand.org/pubs/monographs/MG809.html

Wilsdon, James, Liz Allen, Eleonora Belfiore, Philip Campbell, Stephen Curry, Steven Hill, Richard Jones, Roger Kain, Simon Kerridge, Mike Thelwall, Jane Tinkler, Ian Viney, Paul Wouters, Jude Hill, and Ben Johnson, *The Metric Tide: Report of the Independent Review of the Role of Metrics in Research Assessment and Management*, Higher Education Funding Council for England, 2015. As of August 21, 2018: https://responsiblemetrics.org/the-metric-tide/

Wooding, Steven, Stephen Hanney, Martin Buxton, and Jonathan Grant, *The Returns from Arthritis Research*, Vol. I: *Approach, Analysis and Recommendations*, Santa Monica, Calif.: RAND Corporation, MG-251-ARC, 2004. As of August 21, 2018: https://www.rand.org/pubs/monographs/MG251.html

Wratschko, Katharina, "Empirical Setting: The Pharmaceutical Industry," in Katharina Wratschko, ed., *Strategic Orientation and Alliance Portfolio Configuration: The Interdependence of Strategy and Alliance Portfolio Management*, Gabler, 2009, pp. 87–96.